How to Estimate with
RSMeans Data
Data

How to Estimate with
RSMeans Data

Basic Skills for Building Construction

Fifth Edition

Saleh Mubarak
RSMeans

RS**Means**   WILEY

Copyright © 2020 by John Wiley & Sons, Inc. All rights reserved.

Published by John Wiley & Sons, Inc., Hoboken, New Jersey.

Published simultaneously in Canada.

Cover Design: Wiley
Cover Images: 3D architecture abstract © nadla/iStockphoto; Store Building Exterior © Tony Tremblay/iStockphoto; Steel Construction © SimplyCreativePhotography/iStockphoto; Modern Office Building © Tony Tremblay/iStockphoto; Construction Workers © shotbydave/iStockphoto; Home Under Construction © bbourdages/iStockphoto

Library of Congress Cataloging-in-Publication Data

Names: Mubarak, Saleh A. (Saleh Altayeb) | R.S. Means Company.
Title: How to estimate with RSMeans data : basic skills for building construction / RSMeans, Saleh Mubarak.
Description: Fifth edition. | Hoboken, New Jersey : John Wiley & Sons Inc., [2016] | Includes indexes.
Identifiers: LCCN 2016018464| ISBN 9781118977965 (pbk. : acid-free paper) | ISBN 9781118977972 (epub)
Subjects: LCSH: Building--Estimates.
Classification: LCC TH435 .M79 2016 | DDC 692/.5--dc23 LC record available at https://lccn.loc.gov/2016018464

978-1-118-97796-5

Printed in the United States of America

SKY10059701_111023

Contents

Preface

This fifth edition is a continuation of the success of this book. Success can never come as a coincidence or by luck. It comes only through planned, hard, and intelligent work.

Professional estimators quantify the needed resources—materials, labor, and equipment—required by the scope of a project, and then price these items. This is a two-phase process that includes quantity takeoff and cost estimating. To complete the quantity takeoff, the estimator examines plans and specifications to determine total quantities of materials required, as well as labor and equipment. During the cost estimating phase, the estimator examines the direct costs of installed materials and equipment, labor rates, construction equipment and tool costs, and indirect expenses, such as overhead and profit. Inflation and market conditions are additional factors to consider. The estimator needs also to be familiar with the contract, especially the sections relevant to or impacting the cost.

Special problem-solving skills are required to obtain an accurate estimate. No matter what source is used, construction cost data are rarely available in the perfect format for a particular estimate. Data must often be adapted in some way, such as changing the number of units, the location, production rates, or the type of labor. Frequently, there is "math and more" to be done beyond what is required to produce the quantity takeoff, such as converting units of measure, adjusting for overtime, allowing for difficult access to the site, or factoring in other special considerations. Time-cost trade-off is another important consideration as owners and contractors need in some situations to accelerate the project, which has direct and perhaps complicated impact on the total cost.

This book provides information about how the costs in RSMeans Building Construction Cost Data (BCCD) are developed and presented. It also provides numerous sample problems that show how to apply this cost information. Following these guidelines will enable you to use the BCCD "to the max," creating a detailed estimate, made more accurate by utilizing the full capabilities of the data.

There are many changes in the fifth edition. Chapter 3, "Cost Estimating: An Introduction" has been expanded, adding new sections. I found this chapter particularly important for those who want to get an idea on construction cost estimating without reading an entire book on the subject. More examples and exercises were added. The CSI MasterFormat has been updated according to the 2016 version, which is periodically being updated by the Construction Specifications Institute (CSI) since the major update in 2004 that took the number of divisions from 16 to 50. Chapter 4, "General Requirements," was expanded and moved to the front of the book.

The book now is published by Wiley, an international leader in publishing scientific and professional books. RSMeans is still involved with the book, particularly in updating the materials related to the online estimating (construction cost estimating database and software).[1]

The answers and solutions to the exercises were rearranged on a companion website (www.wiley.com/go/constructionestim5e) where users of the book will be given access. The solutions to Exercises—Set B will be available online only to instructors.

Included with this workbook is an access to the RSMeans Online Estimating, the electronic version of RSMeans Building Construction Cost Data. RSMeans Online Estimating includes the capability to create your own cost list estimates within the program, or to download these estimate directly to a spreadsheet. Users can practice their skills in creating a complete construction estimate using the building plans for a residential and a light commercial structure (provided online).

All numbers in the examples and exercises in this book are based on the 2017 RSMeans BCCD (Building Construction Cost Database).

This book focuses on solution techniques for the various types of estimating problems and on using RSMeans Online Estimating to create a spreadsheet estimate. Theoretical explanations of the various estimating techniques are beyond the scope of this publication.

As we improve in each new edition, we are striving for perfection, which humans can never reach. To me, this is good news because it means there is always room for improvement. This is what motivates us to keep improving with no limitation or ceiling. I hope all users of this book—instructors, students, professionals, and others—to communicate with me or the publisher for any idea or correction that can improve this book. The author can be reached at the email address cpmxpert@gmail.com.

---

1. In the past four editions, RS Means provided with this book a CD containing construction cost estimating database and a software called CostWorks. Although the CD is still available for purchase separately, this fifth edition of the book uses the online estimating instead.

Acknowledgments

This fifth edition comes on the heels of the fourth edition, when this book was published by Wiley, an international leader in publishing. RSMeans, the leading company in construction cost estimating databases, in still a partner and contributor to the book, including the construction cost estimating database and software. Both organizations have been superb in support and service. My experience with RS Means goes back to 1986 when I used their Building Construction Cost Data (BCCD) book as a graduate student. I have used it again as a professor since 1990.

I would like to thank the engineering team at Gordian for their help with this edition. I would also like to thank the team at Wiley for great support. Wiley took care of my book *Construction Project Scheduling and Control* and did a great job. They are the world's experts in publishing and marketing scientific books, and I am glad they are publishing this book as well.

I must also recognize the contribution of Tom Bledsaw, currently with Draper, Inc. and formerly with ITT educational Services as the national chair, School of Drafting and Design, and Harold Grimes, the department chair of construction management and general education at Redstone College, as reviewers of this edition.

Finally, I owe a lot of gratitude to the numerous friends and colleagues who passed their comments on the book to me. As humans, we are far from perfection, but I take this as a motivator: there is always room for improvement.

Introduction

RSMeans data from Gordian provides accurate and up-to-date cost information to help owners, developers, architects, engineers, contractors and others carefully and precisely project and control the cost of both new building construction and renovation projects.

This book is based on the RSMeans Building Construction Costs Database, BCCD, which has been printed in books for over 80 years and is now available online. Along with the BCCD online database, RSMeans has provided a cost estimating software. We will refer to it in the book as the RSMeans Online Estimating. It offers a single line item, assemblies, and square foot estimating programs. Users of the book will be given free access to limited use as a supplement to this book.

Instructional Information

First, the user needs to register. Once registration is complete, user can long in to the site: www.rsmeansonline.com.

The user will be directed to the "Welcome to RSMeans Online!" page, where he/she needs to set own preferences (Figure 00.1).

Figure 00.1

Complete the choice of preferences and then click "Save & Continue". This screen will appear every time the user logs in unless the "Display Cost Data Preferences at start-up" is unchecked.

On the main page, we have the main menu with these options:

1. Search Data: Display of the database, cost line items or assemblies
2. Manage Estimates: Manages the estimates you created
3. Square Foot Estimator: A special program for conceptual estimating with more than 100 commercial and residential models available
4. Life Cycle Cost: An option for improving the long-term performance of buildings and gauge installed costs versus long-term facility maintenance costs
5. Cost Alerts and Trends: An option to receive notifications and track cost trends for the materials, labor, and equipment
6. Reference Items: Supplemental information such as a list of abbreviations, city cost index, crews, labor rates, references (notes that relate to cost line items), estimating tips, dictionary (for cost-related terms), a video tutorial, and student edition materials
7. My Favorites: A special database for items and assemblies you choose as favorite so it will be easy to reuse later

You can always click on the green button "Guide Me" on the upper right-hand side for valuable help lessons.

More explanation on creating and managing estimates through examples in the following chapters.

A Cautionary Note: Numerical Rounding and Mathematical Judgment

Construction cost estimating is not an exact science. It depends on many uncertain factors (labor productivity, price escalation, and so forth) that make absolute accuracy impossible. It is a prediction of future expenses. Assumptions will have to be made about waste factors, contingency costs, takeoff techniques, and many other unknown or uncertain factors. Given the same set of plans and specifications, several estimators will come up with different project totals, all of which will probably differ from the final project cost. The good estimator is the one who gets his estimates consistently close to the actual cost.

Mathematical Intuition

Scholars differ and argue on the definition of *mathematical intuition* and what factors play in measuring or increasing it. In the context of construction cost estimating, we can simply state that a cost estimator must possess a minimum level of mathematical intuition to enable him or her to make good common sense judgments on numbers and to judge whether a number is too high or too low. Such intuition is essential to avoid major mistakes that may lead to financial losses and other negative consequences. Cost estimators with good mathematical intuition also can provide, in most cases, a ballpark figure for the cost of a proposed project without sophisticated methods and tools.

Although scholars may argue, again, on how much of this mathematical intuition is inherited and how much is acquired, there is no question that any human being can enhance it by learning a few simple techniques and continuous practice. It is just like any other mental and physical power that humans possess; it increases—or at least is maintained—by practice, and decreases by neglect and lack of practice. This point is becoming increasingly important as we have entered the digital age and accumulated plenty of electronic gadgets. Technological advancements and inventions continue day after day with no end (or even a slowdown) in sight.

Our increasing dependence on such gadgets is leaving a negative effect on many talents such as the mathematical intuition. For example, there is no question that the quality of the average human's handwriting has declined because of the overwhelming use of computer and other electronic devices' keyboards. People now depend more on their cell phones rather their own memory to store telephone numbers and other information. New technologies have also automated many processes such as structural analysis and design, medical diagnosis, and automotive mechanical and electrical diagnosis. As wonderful as it seems to many people, this should trigger an alarm: many people are losing their professional intuition and analytical capability. They are becoming too dependent on technologies to the point they cannot function or perform simple tasks without their electronic devices. Computers and other electronic gadgets are wonderful tools that can and do help tremendously, but they should never be a replacement for the human intelligence, thinking, and creativity.

To the cost estimator, there are simple exercises that can help build or at least maintain this intuition, such as calculating the value of the groceries or other commodities purchased from a store, including any percent discount and sales tax, and then comparing this approximate total to the cashier's total. One can calculate or estimate the monthly payment on a purchased car and compare it with the amount provided by the salesperson. Practice estimating the height of a high-rise building (in feet or meters, or number of floors), the number of bricks in a pallet or group of pallets of bricks, or the number of openings (doors and windows) in a building. It is always a good idea to do quick and approximate mental math and then compare the answer to the one produced by the computer, calculator, or other devices. When the two answers are significantly different, you might discover that the other answer (the supposedly accurate one) is wrong, either through human input error or a software flaw.

Some Helpful Suggestions

Rounding numbers must be done systematically and with care to avoid the introduction of significant errors. It is recommended when performing calculations that you enter dimensions without rounding, especially those to be multiplied by a large quantity. The amount of error in rounding depends on the number(s) the rounded number is multiplied by.

For example, assume an elevated concrete slab is 211'-11" long, 120'-0" wide, and 7.5" thick. The volume is 588.66 CY. If we rounded the length to 212', the volume would be 588.89 CY, an error of 0.23 CY. However, if we rounded the thickness to 8" instead of 7.5", the volume would be 627.90 CY, a whopping 39.24 CY error.

The explanation is easy: the first error represents a 1" × 120' × 7.5" strip (two small and one large dimensions). The second error represents a 211'-11" × 120' × 0.5" strip (one small and two large dimensions). It is important to be careful with such practices, and avoid rounding in early stages of the estimate.

The estimator should have a sense of the size of the error introduced by rounding to ensure that it will not significantly affect the total estimate. Mathematical intuition and good common sense judgment are a must for a good estimator. As one estimator said, "While the price of one item may be too high or too low, the overall estimate should be pretty accurate."

Be careful when using manual or electronic tools for measuring dimensions. The results produced by rolling pens, digitizing boards, and other tools vary by device and user. Again, the estimator must use common sense judgment to make sure no unmanageable error is introduced into the estimate.

If using a handheld calculator, use one with ordinary fractions capability (*b/c*), so you can enter 8" as 8/12 ft, rather than the decimal fraction 0.67. This eliminates the introduction of a rounding error. Follow the same concept when using Excel.

Avoid false accuracy. As cost estimating is a *prediction* of future expenses, final answers should be rounded to a reasonable degree. As a rule of thumb, a figure with four significant digits is an acceptable accuracy. It would be ridiculous to estimate the total cost of a construction project as $2,148,387.23. This is *false accuracy* because it gives the reader a feeling that this number is very accurate, while in fact it is not. If it refers to *actual* expenses, the previous number may be true and accurate. The following are some examples:

Estimate	Estimate Rounded
$122,778.34	$122,800 or 123,000
$367,289.45	$367,000
$2,446,983	$2,447,000
$53,674,294.55	$53,670,000
$453,681,302.88	$453,700,000

In any mathematical operation, the highest level of accuracy for the answer is the same as the accuracy of the least accurate number of that operation. For example, consider:

$$A = B + C \times (D/E) - F$$

where *B, C, D, E,* and *F* are all real numbers.

The highest accuracy for *A* is the same as the least accuracy for *B, C, D, E,* or *F.* Note that when exact numbers are used, they have a perfect accuracy (or infinite number of significant digits). For example, if we are calculating the volume, in cubic yards, of a concrete footing that measures 3′-4″ by 3′-4″ by 1′-4″, the answer would be:

$$\text{Volume (in CY)} = \frac{\text{Length (in feet)} \times \text{Width (in feet)} \times \text{Depth (in feet)}}{27 \text{ CF/CY}}$$

$$= \frac{(40/12 \times 40/12 \times 16/12)}{27 \text{ feet}^3/\text{yard}^3} = 0.548696845 \text{ CY}$$

The answer can be rounded to any number of significant digits we desire, such as 0.55, 0.549, 0.5487, and so on.

Suppose volume is written as:

$$\text{Volume} = \frac{(3.33' \times 3.33' \times 1.33)}{(27 \text{ feet}^3/\text{yard}^3)} = 0.546 \text{ cubic yards}$$

The answer cannot have more than three significant digits because the least accurate number used in the operation had only three significant digits.

RSMeans uses the following rounding standards:

Prices From		To		Rounded to nearest
$	0.01	$	5.00	$0.01
	5.01		20.00	0.05
	20.01		100.00	1.00
	100.01		1,000.00	5.00
	1,000.01		10,000.00	25.00
	10,000.01		10,000.00	100.00
	50,000.01		Up	500.00

Use educated common sense judgment. Human errors and equipment malfunctions are always possible. For example, when entering a number in a calculator or a computer keyboard, you may intend to press the 8 key, but the key got stuck and multiple of 8s were displayed instead, or nothing at all. To minimize such errors, follow these three rules:

1. Always keep an eye on the computer screen or calculator display to make sure it matches the entered number.

2. Apply common sense judgment to the answer. For example, if you are calculating the cost of erecting wood joists for one floor in a 2,000 SF house, and the answer was too high (e.g., $459,000) or too low (e.g., $570), you know there is something wrong. This judgment depends on the estimator's experience and construction common sense judgment.

3. Be well-organized and maintain a calculation audit trail. In case of a review or a suspected error, it should be easy for you or anyone else to follow your work, step by step.

Chapter 1
Basic Calculations

RSMeans Building Construction Cost Data (BCCD) is the most widely used reference book for estimating construction costs in the United States and Canada. The costs for each construction item are broken down into the components of material, labor, equipment, and overhead and profit. The book also contains square foot costs by project type. The square foot cost data must be adjusted to fit the specific location, size, and conditions of a particular project. RSMeans Online Estimating and Cost Works[1] are electronic versions of the *BCCD* and contain all the same information plus additional features—including the ability to adjust all cost figures by a specific location factor, apply quantities to line items, and export cost data to a spreadsheet.

RSMeans Cost Data Format

The RSMeans Unit Price Line

All RSMeans unit price (UP) data are presented in the same basic format (Figure 1.1).

Each line in the RSMeans database contains information unique to that line: a specific 12-digit number address, detailed description, crew, daily output of the task using the noted crew, labor hours for the task using the specific crew, and a unit of measurement. Also included are the unit

★	⚡	Line Number	⬩	✎	Description	Unit	Crew	Daily Output	Labor Hours	Bare Material	Bare Labor	Bare Equipment	Bare Total	Total O&P
		033053.40			Concrete In Place									
☆	⚖	033053400010			CONCRETE IN PLACE									
☆	⚖	033053400020			Including forms (4 uses), Grade 60 rebar, concrete (Portland cement									
☆	⚖	033053400050			Type I), placement and finishing unless otherwise indicated									
★	⚖	033053400300			Beams (3500 psi), 5 kip per L.F., 10' span	C.Y.	C14A	15.62	12.804	330.00	635.00	53.50	1018.50	1389.00
☆	⚖	033053400350			25' span	C.Y.	C14A	18.55	10.782	345.00	535.00	45.00	925.00	1229.50

Figure 1.1

---

1. An RSMeans product for construction cost estimating that is provided on a CD and does not require Internet access.

bare cost (material, labor, equipment, and total) and the total unit cost, including overhead and profit.

Unit price information is presented according to the 50 divisions of the Construction Specifications Institute (CSI) MasterFormat 2016. These divisions are divided into major subdivisions and then into subsections of similar items. Within each subsection, items are arranged alphabetically by type. Each line item is unique.

Address Number

The address number of the line item shown here can be read as three parts:

03 30 Cast-In-Place Concrete is the Level Two, CSI MasterFormat subdivision. The first two digits of that number represent the Level One, MasterFormat division. (For this item, it is Division 3, or 03, which is Concrete.)

03 30 53 Miscellaneous Cast-In-Place Concrete is the Level Three subdivision.

Concrete In Place is the Level Four, RSMeans major classification. It appears in the RSMeans book as a line 03 30 53. 40, while it appears in RSMeans online estimating as just the number 40 in the extreme left and right columns.

0350 (first column from left in the RSMeans book and second column from left in RSMeans online estimating) is the RSMeans individual line number.

Note that while MasterFormat used spaces and a period in the number for better reading (03 30 53.40 0350), no spaces or period should be used when searching for an item in RSMeans online estimating software.

Description

The column to the right of the Line Number, for example 033053400350, contains a detailed description of the item. For a full description of an item, one must read up through the subsection, including all descriptive information that appears on lines above and to the left of the selected item.

Thus, the complete description for the item in line 03 30 53.40 0350 is "Concrete in place including forms (4 uses), Grade 60 rebar, concrete (Portland cement Type I), placement and finishing unless otherwise indicated, Beams (3500 psi), 5 kip per LF, 25' span." An easier way to read description is to click on the Line Number, as shown in Figure 1.2.

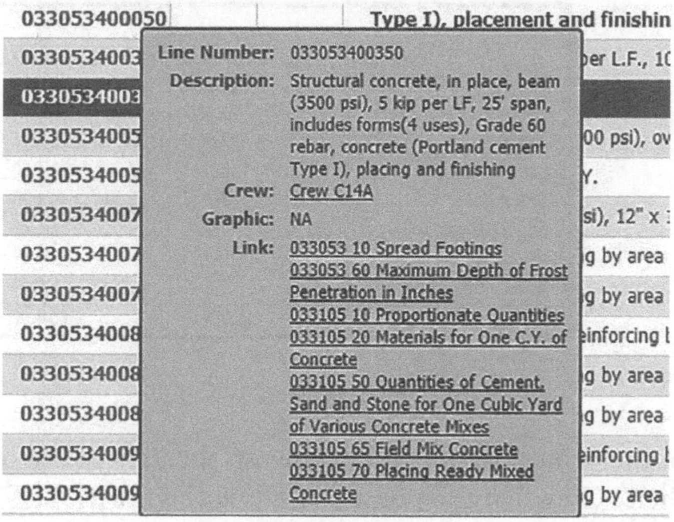

Figure 1.2

Note that the wording here may not match the example in the previous paragraph exactly, but it is the same content.

In the RSMeans cost estimating database print books, the description box of some items may include an illustrative sketch or a reference box. The reference box next to the item indicates an RSMeans reference number or assembly. Reference numbers and assemblies contain detailed information that may be helpful to the estimator. Several reference boxes indicate that all these notes apply to all the items in that section. For example, R033053-10, R033053-60, R033105–10, R033105–20, R033105–50, R033105–65, and R033105–70 apply to items 03 30 53.40 0010 through 03 30 53.40 7050. They show in the pop-up window in Figure 1.2 under the title Link. View the note by clicking on the Reference number; a new tab opens to show the note in pdf format. It can be downloaded and saved or printed. In the same pop-up window are other categories: Crew (to be explained next) and Graphics, which may include a clickable file name. Most likely, this graphics is G.pdf, HANDICAP.pdf, or another file. The G.pdf is usually accompanied by a green leaf in the main spread sheet, indicating that this item was identified by the RSMeans engineering staff to be environmentally responsible and/or resource-efficient. The HANDICAP.pdf simply shows the handicap logo. Other items may have a graphic illustration.

Crew

In the database, the term *crew* refers to a unique grouping of workers and equipment, identified by letter and number. The crew on each line includes the labor trade or trades and equipment required to efficiently perform the indicated task. Crew details are shown in the reference section at the back

Crew C-14A	Hr.	Daily	Hr.	Daily	Bare Costs	Incl. O&P
1 Carpenter Foreman (outside)	$51.25	$410.00	$78.50	$628.00	$49.40	$75.64
16 Carpenters	49.25	6304.00	75.45	9657.60		
4 Rodmen (reinf.)	54.30	1737.60	83.70	2678.40		
2 Laborers	39.15	626.40	60.00	960.00		
1 Cement Finisher	46.80	374.40	69.60	556.80		
1 Equip. Oper. (medium)	53.55	428.40	81.00	648.00		
1 Gas Engine Vibrator		28.25		31.07		
1 Concrete Pump (Small)		806.40		887.04	4.17	4.59
200 L.H., Daily Totals		$10715.45		$16046.92	$53.58	$80.23

Figure 1.3

of the print book or by clicking on the item in the online database and then "Crew" in the pop-up window (Figure 1.3).

The opened tab shows all crews in pdf format, listed in alpha-numeric order. Another way to obtain the crew list is to click "Reference Items" on the top bar menu, and choose Crews. The "Reference Items" page gives a host of other useful information.

Crew labor hours are shown in the lower-left corner of the crew box. In the case of Crew C14A, the total is: 200 LH Daily Totals. This figure represents the total labor hours worked by the 25 members of Crew C14A in a normal eight-hour workday.

If the task is done by one type of laborer with no equipment (e.g., the crew is composed of one trade only), there will be no crew ID. Instead, there will be an abbreviation for that particular labor trade and the number of workers. For example, most items in section 03 21 11 60 0360 are done by rodmen, abbreviated as Rodm. Hence, Rodm appears in the crew column. The labor trade table lists the abbreviations and their detailed cost information. The table can be seen on the inside back cover of the *BCCD* book or it can be obtained online by choosing Labor Rates in the "Reference Items."

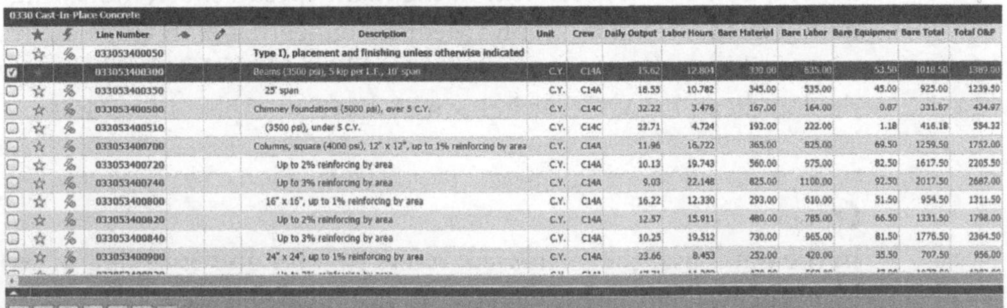
0330 Cast-In-Place Concrete

Line Number	Description	Unit	Crew	Daily Output	Labor Hours	Bare Material	Bare Labor	Bare Equipmen	Bare Total	Total O&P
033053400050	Type I), placement and finishing unless otherwise indicated									
033053400300	Beams (3500 psi), 5 kip per L.F., 10' span	C.Y.	C14A	15.62	12.804	330.00	635.00	53.50	1018.50	1389.00
033053400350	25' span	C.Y.	C14A	18.55	10.782	345.00	535.00	45.00	925.00	1239.50
033053400500	Chimney foundations (5000 psi), over 5 C.Y.	C.Y.	C14C	32.22	3.476	167.00	164.00	0.87	331.87	434.97
033053400510	(3500 psi), under 5 C.Y.	C.Y.	C14C	23.71	4.724	193.00	222.00	1.18	416.18	554.32
033053400700	Columns, square (4000 psi), 12" x 12", up to 1% reinforcing by area	C.Y.	C14A	11.96	16.722	365.00	825.00	69.50	1259.50	1752.00
033053400720	Up to 2% reinforcing by area	C.Y.	C14A	10.13	19.743	560.00	975.00	82.50	1617.50	2205.50
033053400740	Up to 3% reinforcing by area	C.Y.	C14A	9.03	22.148	825.00	1100.00	92.50	2017.50	2687.00
033053400800	16" x 16", up to 1% reinforcing by area	C.Y.	C14A	16.22	12.330	293.00	610.00	51.50	954.50	1311.50
033053400820	Up to 2% reinforcing by area	C.Y.	C14A	12.57	15.911	480.00	785.00	66.50	1331.50	1798.00
033053400840	Up to 3% reinforcing by area	C.Y.	C14A	10.25	19.512	730.00	965.00	81.50	1776.50	2364.50
033053400900	24" x 24", up to 1% reinforcing by area	C.Y.	C14A	23.66	8.453	252.00	420.00	35.50	707.50	956.00

Figure 1.4

Most columns can be resized by grabbing and dragging the divider lines at the tops of the columns.

Costs in the crew details box are itemized in three ways:

1. Bare costs
2. Including subs O&P (overhead and profit)
3. Cost per labor hour

The bare cost is based on the wages displayed in column A of the table "Installing Contractor's Overhead and Profit" (located on the inside back cover of the printed book, or in the Crew Information (as show in Figure 1.3). The cost "Including Subs O&P" is based on the labor wages with add-ons, displayed in columns H and I of the same table. Equipment cost, including O&P, is calculated by adding 10 percent to the bare equipment cost. The cost per labor hour is based on labor and equipment cost, divided by total labor hours.

In our example, total daily labor bare cost of crew C14A is

$$\$410.00 + \$6,304.00 + \$1,737.60 + \$626.40 + \$374.40 + \$428.40 = \$9,880.80/day$$

If this number is divided by 200 labor hours per day, we'll get $49.40 per average labor hour. If we repeat these steps with labor wages including O&P, we'll get $15,128.80/200 = $75.64 per average labor hour.

For equipment, total daily bare cost is $28.25 + $806.40 = $834.65, or $4.17 average per labor hour, or $918.11/200 = $4.59 average per labor hour, including O&P. This is a hypothetical number that represents the average equipment cost per labor hour if the equipment costs are spread evenly among all workers.

When a crew contains a 0.5 or 0.25 worker, it means the worker is working a half day (4 hours) or a quarter of a day (2 hours) during a normal workday.

Daily Output

The number of units of a defined task that a designated crew will produce in one eight-hour workday is referred to as the *daily output*. Daily output represents an average figure, which will vary with job conditions. Daily output is measured in the units specified in the unit column. For line 03 30 53.40 0350, the output is 18.55 CY (cubic yards) per day.

Labor Hours

This number represents the total number of labor hours it takes to produce one unit of this task using the specified crew. Labor hours per unit is calculated by dividing the crew labor hours (found in the crew detail) by the daily output:

$$\text{Labor hours/Unit} = \frac{\text{Daily crew labor hours}}{\text{Daily output}}$$

$$\text{Labor hours/CY} = \frac{200\,\text{Labor hours/day}}{18.55\,\text{CY/day}} = 10.782\,\text{Labor hours/CY}$$

This basic relationship of crew labor hours and productivity can be used to calculate labor hours for crews of different composition. It can also be used to calculate the length of time it will take to perform this task with crews of differing composition.

One important observation about the two terms representing productivity—namely, units/day and labor hours/unit—is the adjustment needed when productivity changes. There is an inverse proportionality between the two terms. For example, if productivity decreases, then units/day decreases while labor hours/unit increases, and vice versa.

Unit

In most cases, the unit is self-explanatory, such as CY (cubic yards), SF (square feet), or Ea (each). In some cases, the user might see an unusual unit, such as SFCA (square foot contact area), SQ (square = 100 SF), Cwt (100 pounds), or VLF (vertical linear feet). Refer to the abbreviations list in the reference section of the *BCCD*, or use "Reference Items" in RSMeans Online Estimating.

Bare Costs

This category has four columns: Materials, Labor, Equipment, and Total. The numbers here represent the contractor's direct (bare) cost. They do not include any overhead, subcontractors' markups, or profit.

Total Incl. O&P

This column represents the sum of the bare material cost plus 10 percent for profit, the bare labor cost plus labor burden and 10 percent for profit, and the bare equipment cost plus 10 percent for profit. This figure represents the amount the installing contractor may charge to a general contractor or owner.

Productivity and Activity Duration

Labor cost is, without a doubt, the single most unpredictable expense in a construction project. Labor cost can be calculated in at least two ways, as shown in equations (1) and (2):

(1) $\text{Labor cost (\$)} = \text{Activity duration (days)} \times \text{Crew cost (\$ / day)}$

or

(2) $\text{Labor cost (\$)} = \text{Units to be completed} \times \text{Labor cost (\$ / unit)}$

Note: The unit of time used in activity duration could be hours, weeks, or any other unit of time. The crew cost then must be in dollars per the same time unit.

Both methods require knowing the productivity—that is, the total number of units produced by the unit of time. If we define productivity as a crew's daily output (units/day), then:

$$\text{Activity duration} = \frac{\text{Units to be completed (size of job)}}{\text{Daily output}}$$

and

$$\text{Labor unit cost (\$/unit)} = \frac{\text{Crew labor cost (\$/day)}}{\text{Daily output (units/day)}}$$

Then equations (1) and (2) can be rewritten as one equation:

(3) $\text{Labor cost(\$)} = \dfrac{\text{Units to be completed}}{\text{Daily output}} \times \text{Crew labor cost (\$/day)}$

RSMeans productivity figures assume average conditions. Productivity involves two indicators: daily output and labor hours (per unit). For any given task:

$$\text{Labor hours} = \frac{\text{Crew labor hours}}{\text{Daily output}}$$

Example 1

For line 31 23 16.42 1600, an excavation crew (B10T) consists of one equipment operator, 0.5 laborer (4 hours/day), and a front-end loader. It has an output of 800 CY/day. What are the labor hours/CY?

Solution

The total crew labor hours/day for crew B10T is $8 + 4 = 12$ labor hours.

$$\text{Labor hours / CY} = \frac{12 \text{ labor hours / day}}{800 \text{ CY / day}} = 0.015 \text{ labor hours / CY}$$

Example 2

A crew of three bricklayers and two helpers laid 1,200 bricks in six hours. Calculate the daily output and the labor hours/M bricks.[1]

Solution

Hourly output = 1,200 bricks/6 hours = 200 bricks/hour

$$\text{Daily output} = 200 \text{ bricks/hour} \times 8 \text{ hours/day}$$
$$= 1,600 \text{ bricks}$$
$$= 1.6 \text{ M bricks} (M = 1000)$$
$$\text{Crew labor hours/day} = 5 \text{ workers} \times 8 \text{ hours/day}$$
$$= 40 \text{ labor hours}$$
$$\text{Labor hours} = 40 \text{ labor hours/1.6 M bricks}$$
$$\text{Labor hours/M bricks} = 25 \text{ labor hours/M bricks}$$

or:

In 6 hours, the crew puts out 30 labor hours (5 men × 6 hours)

$$30 \text{ labor hours} = 1,200 \text{ bricks}$$
$$\text{Labor hours/brick} = 30 \text{ labor hours/1,200}$$
$$= 0.025 \text{ labor hours/brick}$$
$$= 25 \text{ labor hours/M bricks}$$

---

[1] M bricks means 1,000 bricks; in Roman numerals, the letter "M" represents 1,000.

If productivity increases (or decreases) by a certain percentage, the daily output will also increase (or decrease) by the same percentage. The labor hours per unit will go the opposite way—that is, they will decrease (or increase) by the same percentage. To understand this, let's consider the following scenario.

Example 3

The productivity of the crew in Example 1 dropped by 20 percent due to some adverse conditions. What would be the new daily output and labor hours/CY?

Solution

$$\text{New output} = 800\,\text{CY/day} \times (100\% - 20\%) = 800 \times (0.80) = 640\,\text{CY/day}$$
$$\text{New labor hours} = 0.015\,\text{LH/CY}/(100\% - 20\%) = 0.015/0.8$$
$$= 0.01875\,\text{LH/CY}$$

Note that dividing by $(100\% - 20\%)$ is not equal to multiplying by $(100\% + 20\%)$. In case of a productivity drop, always use $(100\% - x\%)$. In case of productivity gains, always use $(100\% + x\%)$, where x is the productivity change.

Example 4

Calculate the estimated duration for excavating 10,000 CY of earth using Crew B10T of Example 1.

Solution

$$\text{Duration} = \frac{10,000\,\text{CY}}{800\,\text{CY/day}} = 12.5\,\text{days} \approx 13\,\text{days}$$

Duration varies significantly, not only with productivity but also by crew size or composition, or the number of crews used. Duration has an inverse relationship with productivity—that is, when productivity increases, duration decreases, and vice versa.

The estimator must exercise caution in calculating the duration of an activity. Duration is important for scheduling and calculating the expected finish time (for that activity and the entire project) as well as for estimating general conditions and overhead cost (rent, staff salary, utilities, etc.) of the job.

Equipment Costs

Equipment for each task is shown in the crew detail. Crew equipment cost consists of equipment rental costs plus equipment operating costs.

$$\text{Equipment cost/unit} = \frac{\text{Crew equipment cost/day}}{\text{Daily output}}$$

For line 03 30 53.40 0350:

$$\text{Equipment cost} = \$45/CY$$

Using Crew C-14A information:

$$\text{Equipment cost} = (\$28.25 + \$806.40)/18.55CY/day$$

$$= \$45/CY$$

RSMeans cost data for equipment are based on two assumptions:

1. Late-model, high-quality machines in excellent working condition, rented from equipment dealers[2].

2. Weekly rental:

$$\text{Crew equip. cost/day} = \text{Hourly oper. cost} \times 8 \text{ hours/day}$$
$$+ \frac{\text{Rent per week}}{5 \text{ workdays per week}}$$

For the gas engine vibrator in Crew C14A, refer to line 01 54 33.10 3000.

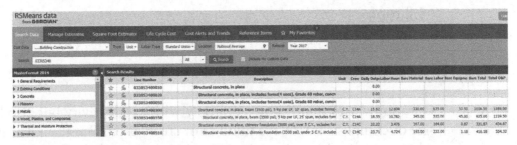

Figure 1.5

For this equipment, the hourly operating cost is $2.38 and the rent per week is $46. The crew equipment cost per day is:

$$\text{Crew equip. cost/day} = \frac{\$2.38}{\text{hour}} \times \frac{8 \text{ hours}}{\text{day}} + \frac{\$46 \text{ per week}}{5 \text{ workdays per week}}$$

---

2. See RSMeans Reference Number R015433-10.

Crew equip. cost = $28.24/day (in the crew cost, it shows as $28.25 because of different rounding process).

This crew equipment cost per day (shown in the far-right-hand column in the equipment section in the print book only) is the figure used in the crew detail. Construction equipment is often rented for periods other than one week or may be owned. Variations in times for equipment rental will be discussed further in Chapter 13.

City Cost Indexes and Location Factors

The costs shown in the unit price section of the *BCCD* represent US national average prices and are given in US dollars. The *national average* is an average of costs in 30 major US cities. National average costs should be adjusted to local costs when performing an actual estimate. City cost indexes for 731 US and Canadian cities are available in the reference section of the *BCCD*. The 30-city national average is given a value of 100, and each of the 731 index figures is shown as a value relative to this figure. They are broken down not only by city but also by the major CSI division.

In the *BCCD* print book, the location factors list the same factors for 731 cities with the addition of zip codes (but without breakdown by CSI divisions). To determine the cost in a particular location, multiply the cost by the location factor and divide by 100.

$$\text{Cost in City A} = \text{National average cost} \times \frac{\text{Location factor}}{100}$$

All cost numbers in the *BCCD* print book show US national averages, and then the user needs to multiply the cost number by CCI factors. In the Online Estimating, it is adjusted automatically by choosing the Location: A small window will pop up, and the user chooses one of the available cities in the United States or Canada (Figure 1.6).

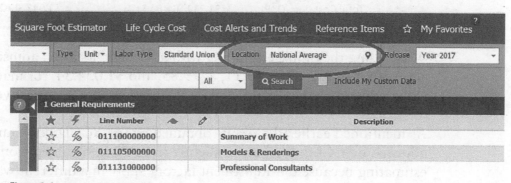

Figure 1.6 Project Location

In the examples in the next chapters, we will be using the numbers from the online estimating, that are already adjusted for location.

Example 5

What is the cost per CY for the concrete beams in RSMeans item 03 30 53.40 0350 in Omaha, Nebraska?

03 30 53.40 Concrete In Place													
★	⚡	Line Number	⚐	✎	Description	Unit	Crew	Daily Output	Labor Hours	Bare Material	Bare Labor	Bare Equipment	Bare Total	Total O&P
☆	⚡	033053400010			CONCRETE IN PLACE									
☆	⚡	033053400020			Including forms (4 uses), Grade 60 rebar, concrete (Portland cement									
☆	⚡	033053400050			Type I), placement and finishing unless otherwise indicated									
★	⚡	033053400300			Beams (3500 psi), 5 kip per L.F., 10' span	C.Y.	C14A	15.62	12.804	330.00	635.00	53.50	1018.50	1389.00
☆	⚡	033053400350			25' span	C.Y.	C14A	18.55	10.782	345.00	535.00	45.00	925.00	1239.50

Figure 1.7

Solution

The bare cost per CY for the concrete beams in line item 03 30 53.40 0350, US national average, is $925, which includes $345, $535, and $45 for materials, labor, and equipment, respectively. The cost including O&P (overhead and profit) is $1,239.50/CY. These numbers represent the national average for that item. City cost index (CCI) numbers are broken down by division and by subdivision for concrete. But since this item combines all concrete subdivisions (formwork, reinforcement, and concrete mix and placement), we will use the CCI factors for Division 03 Concrete for Omaha, Nebraska—90.8 percent for materials, 77.3 percent for installation (labor), and 95.8 percent for equipment, or 84.5 percent for total cost—as a percentage of the national average.

Using these factors we can find the cost per CY in Omaha:

Cost for Materials	$= \$345 \times 90.8\%$	$= \$313.26/CY$
Cost for Labor	$= \$535 \times 77.3\%$	$= \$413.56/CY$
Cost for Equipment	$= \$45 \times 95.8\%$	$= \$43.11/CY$
Total Bare Cost	$= \$335.34 + \$379.95 + \$37.58$	$= \$769.93/CY$

or

	$= \$925 \times 84.5\%$	$= \$781.63/CY$
Total Cost (incl. O&P)	$= \$1,239.50 \times 84.5\%$	$= \$1,047.38/CY$

When using the online estimating with automatic location adjustment, we get: $309.81, $426.93, $43.11, $779.85, and $1,034.54, for materials, labor, equipment, bare total, and total including O&P; respectively.

As mentioned earlier, the results obtained by the *BCCD* book (and adjusted by CCI factors) differ slightly from those obtained from online estimating because the adjustment factors apply in a different way, and because of rounding (Figure 1.8).

DIVISION		NEBRASKA											NEVADA					
		NORFOLK			NORTH PLATTE			OMAHA			VALENTINE			CARSON CITY			ELKO		
		MAT.	INST.	TOTAL	MAT.	INST.	TOTAL	MAT.	INST.	TOTAL	MAT.	INST.	TOTAL	MAT.	INST.	TOTAL	MAT.	INST.	TOTAL
015433	CONTRACTOR EQUIPMENT		95.8	95.8		106.9	106.9		95.8	95.8		99.4	99.4		98.8	98.8		98.8	98.8
0241, 31 - 34	SITE & INFRASTRUCTURE, DEMOLITION	84.0	94.0	90.9	103.8	94.7	97.4	91.3	94.7	93.7	87.2	100.1	96.2	87.7	98.5	95.2	70.1	97.2	89.0
0310	Concrete Forming & Accessories	83.3	74.3	75.5	96.3	70.7	74.1	95.9	75.4	78.2	84.2	55.2	59.2	108.1	80.2	84.0	113.8	77.0	81.9
0320	Concrete Reinforcing	101.0	64.9	82.7	103.2	73.8	88.3	99.5	74.1	86.6	103.8	64.6	83.9	105.9	109.7	107.8	113.7	90.9	102.1
0330	Cast-in-Place Concrete	108.3	60.1	89.9	114.1	62.0	94.3	89.8	79.8	86.0	100.7	55.1	83.4	95.3	82.0	90.2	92.3	75.0	85.7
03	CONCRETE	101.1	68.3	85.9	107.5	69.5	89.9	90.8	77.3	84.5	105.0	58.0	83.2	97.2	86.1	92.1	95.0	78.9	87.5
04	MASONRY	125.1	74.5	93.9	95.3	70.9	80.3	99.5	75.6	84.8	107.4	70.9	84.9	117.1	70.7	88.5	123.8	69.5	90.4
05	METALS	98.4	78.6	92.3	96.9	91.4	95.2	98.5	83.8	94.0	109.5	78.3	99.9	108.2	92.7	103.5	111.8	83.1	103.0
06	WOOD, PLASTICS & COMPOSITES	82.7	75.9	79.0	97.1	71.6	83.1	94.5	75.9	84.3	81.0	51.2	64.7	91.2	80.3	85.2	102.0	77.1	88.3
07	THERMAL & MOISTURE PROTECTION	102.5	78.6	92.6	99.0	77.8	90.2	98.6	80.0	90.9	99.9	74.2	89.2	106.6	79.4	95.3	103.3	73.2	90.8
08	OPENINGS	95.6	69.9	89.8	94.1	69.9	88.6	100.9	75.4	95.1	96.5	56.4	87.5	98.1	80.8	94.2	101.7	72.7	95.1
0920	Plaster & Gypsum Board	94.7	75.8	82.1	94.7	70.9	78.9	101.4	75.8	84.4	96.1	50.4	65.7	98.3	79.7	86.0	101.3	76.4	84.8
0950, 0980	Ceilings & Acoustic Treatment	96.1	75.8	82.6	86.5	70.9	76.1	95.3	75.8	82.3	101.9	50.4	67.7	96.9	79.7	85.5	97.4	76.4	83.4
0960	Flooring	109.0	93.1	104.4	93.9	93.1	93.7	100.4	85.1	96.0	119.8	74.0	106.5	102.7	71.1	93.5	108.0	71.1	97.3
0970, 0990	Wall Finishes & Painting/Coating	132.2	62.7	91.4	87.5	62.7	73.0	105.0	66.1	82.2	154.2	65.0	101.8	97.3	78.1	86.1	96.8	75.7	84.4
09	FINISHES	102.8	76.4	88.4	91.6	73.6	81.7	97.9	76.2	86.0	111.9	58.4	82.6	95.7	78.2	86.1	95.8	75.7	84.8
COVERS	DIVS. 10 - 14, 25, 28, 41, 43, 44, 46	100.0	84.5	96.5	100.0	63.2	91.7	100.0	88.2	97.3	100.0	59.1	90.7	100.0	81.8	95.9	100.0	92.9	98.4
21, 22, 23	FIRE SUPPRESSION, PLUMBING & HVAC	96.2	75.3	86.9	100.0	74.8	88.8	99.9	76.0	89.3	96.0	74.7	86.6	100.0	70.1	86.8	98.4	75.5	88.3
26, 27, 3370	ELECTRICAL, COMMUNICATIONS & UTIL.	90.5	82.3	86.2	93.4	67.3	79.6	98.8	82.9	90.4	90.6	67.3	78.3	104.1	91.0	97.2	101.3	89.3	94.9
MF2014	WEIGHTED AVERAGE	98.6	77.3	89.2	98.2	75.1	88.0	98.0	79.8	90.0	100.9	69.7	87.1	101.6	81.8	92.8	101.5	80.1	92.0

Figure 1.8

In RSMeans online estimating, location factors can be applied by clicking on Location on the Settings or in the estimate. A choice first must be made among: National Average (US), United States, or Canada. If the choice is United States or Canada, a dropdown menu appears to choose the state. Once a state is chosen, a final dropdown menu appears to choose the city (Figure 1.9). The location can be changed at any time, in the database or the estimate. If this happens while an estimate is open, the prices will adjust automatically but the user needs to save the estimate before / when exiting to keep this change.

Localize your data

Location	United States ▼
State/Region	FLORIDA ▼
City	MIAMI (330-332,340) ▼

Cancel | Save & Continue

Figure 1.9

Chapter 2

Spreadsheet Types

A *spreadsheet* is simply a sheet containing rows and columns where the user can enter data and perform mathematical calculations. Estimators use spreadsheets when performing takeoffs, pricing estimates, and creating estimate summaries to add markups to bare costs. Spreadsheets may be manual (paper and pencil) or electronic. They may be commercially purchased or custom-designed for internal use within a company. When used efficiently, they can be powerful tools to assist the estimator in managing large quantities of information.

Manual Spreadsheets

Estimators benefit from organizing their work in a format that is easy to read and review. Due to the variety of construction items, there is no single "best form." The user may develop his or her own form or use a third-party/commercial one. Examples of manual forms specifically for use with some of the examples in this book can be found in Appendix D.

In most formats (see Figure 2.1), the first column contains the description of the item being considered. Columns follow for quantity (how many); dimensions (with subcolumns for length, width, etc.); units; and notes. Other columns are possible for unit price (with possible subcolumns for labor, materials, equipment); extension (Unit price×Quantity); and/or other types of data.

Each estimate form must contain, at the top of the form, a header with identifying information, such as:

1. Name and address of the company
2. Name of estimator and supervisor (Estimated By and Checked By)
3. Name and location of project
4. Date and perhaps time
5. Page number and total number of pages (Page ___ of ___)

6. Source of the information

7. Any other identifying information, such as Revised Estimate No., Waste Factor, Material Discount, and Labor Rate

ESTIMATE SUMMARY				SHEET NO.		
PROJECT				ESTIMATE NO:		
LOCATION		TOTAL AREA/VOLUME		DATE		
ARCHITECT		COST PER S.F./C.F.		NO. OF STORIES		
PRICES BY:		EXTENSIONS BY:		CHECKED BY:		
.DIV.	DESCRIPTION			MATERIAL	LABOR	
1.0	**General Requirements**					
	Insurance, Taxes, Bonds					
	Equipment & Tools					
	Design, Engineering, Supervision					
2.0	**Site Work**					
	Site Preparation, Demolition					
	Earthwork					
	Caissons & Piling					
	Drainage & Utilities					
	Paving & Surfacing					
	Site Improvements, Landscaping					
3.0	**Concrete**					
	Formwork					
	Reinforcing Steel & Mesh					
	Foundations					
	Superstructure					
	Precast Concrete					
4.0	**Masonry**					
	Mortar & Reinforcing					
	Brick, Block, Stonework					
5.0	**Metal**					
	Structural Steel					
	Open-Web Joists					
	Steel Deck					
	Misc. & Ornamental Metals					
	Fasteners, Rough Hardware					
6.0	**Carpentry**					
	Rough					
	Finish					
	Architectural Woodwork					
7.0	**Moisture & Thermal Protection**					
	Water & Dampproofing					
	Insulation & Fireproofing					
	Roofing & Sheet Metal					
	Siding					
	Roof Accessories					
8.0	**Doors, Windows, Glass**					
	Doors & Frames					
	Windows					
	Finish Hardware					
	Glass & Glazing					
	Curtain Wall & Entrances					
	PAGE TOTALS					

Page 1 of 2

Figure 2.1

A complete Estimate Summary form is shown in Appendix D.

Many companies have developed their own forms with the name and logo of the company at the top of each page. The advantage to this is consistency and ease of tracking. The disadvantage is being limited to a specific form.

Electronic Spreadsheets

There are a variety of electronic spreadsheet programs available for personal computers, such as Microsoft Excel, and Quattro-Pro. Specialized cost estimating programs are also available.[1] Consumers can choose from a wide variety of construction cost-estimating software programs based on several criteria: high-end or low-end, Excel-based, type of construction (home builders, commercial, industrial), CAD or BIM compatible, general or specialized subcontractor, and other. All of these programs have a built-in spreadsheet and a database. Some can link to accounting, scheduling, contract management, or other software.

When using an electronic spreadsheet program, it is important to be familiar with the software's specific mathematical functions, including adding and multiplying, as well as more complex mathematical, statistical, logical, and other functions. Excel has become friendlier, and many of its functions appear when the user starts typing the desired function, such as average, median, sum, square root (sqrt), power, sine (sin), log, and so on. The next example shows mathematical operations in Microsoft Excel.

Example 1

Estimate the cost of furnishing and installing:

40 concrete columns, 24″×24″, 15′ high, with average reinforcement, and

40 concrete spread footings, 6′×6′, 24″ deep.

Adjust costs to Miami, Florida.

Solution

A	B	C	D	E	F	G	H	I	J	K	L	M	N	O	P
				Dimensions					Bare Cost					Total	
Item No.	Description	Unit	How Many?	L'	W'	D"	Qty. Ea.	Qty. Total	Unit Material	Unit Labor	Unit Equipment	Unit Total	Unit Total Incl O&P	Total Bare Cost	Total Incl. O&P
1	Structural concrete, in place, column (4000 psi), square, avg reinforcing, 24" x 24", includes forms(4 uses), reinforcing steel, concrete, placing and finishing	C.Y.	40	15	2	24	2.22	88.89	$376.68	$357.28	$ 45.40	$779.36	$1,011.82	$69,276	$89,940
2	Structural concrete, in place, spread footing (3000 psi), over 5 C.Y., includes forms, reinforcing steel, concrete, placing and finishing	C.Y.	40	6	6	24	2.67	106.67	$176.08	$ 44.98	$ 0.36	$221.42	$263.78	$23,618	$28,137
													Total	$92,895	$118,076

Figure 2.2 This Excel spreadsheet calculates the quantities and then applies units prices from RS Means online estimating, already adjusted for Miami, FL, to calculate total cost.

---

1. IntelliBid by ConEst Software Systems, B2W Estimate by B2W Software, RIB America, CoConstruct, The EDGE by The Estimating Edge, STACK by Stack Construction Technologies, ProContractor Estimating by ViewPoint, Sage Estimating, ProEst Estimating, McCormick Estimating Software, WinEst by WinEstimator, Inc., CostOS by Nomitech, Aspen Capital Cost Estimator by Aspen Technologies (for approximate estimates).

Unit Estimate Summary By Subdivision

Line Number	Long Description	Quantity	Extended Total	Extended Total OP
03				
033053400920	Structural concrete, in place, column (4000 psi), square, up to 2% reinforcing by area, 24" x 24", includes forms(4 uses), Grade 60 rebar, concrete (Portland cement Type I), placing and finishing	88.89	$69,277.31	$89,940.68
033053403850	Structural concrete, in place, spread footing (3000 psi), over 5 C.Y., includes forms(4 uses), Grade 60 rebar, concrete (Portland cement Type I), placing and finishing	106.67	$23,617.80	$28,137.41
		195.56	$92,895.11	$118,078.09

Figure 2.3   Same result as in Figure 2.2, produced by RS Means online estimating, Advanced Report

Assume the concrete column item is in row 3, the footing is in row 4 of the spreadsheet (Microsoft Excel), and the columns in the spreadsheet have column designations the same as those designated in the example. Some of the operations performed above (Figures 2.2 and 2.3) are:

Cell	Math Formula	Excel Formula
H3	E * F * G	= E3 * F3 * G3/12/27
I3	D * H	= D3 * H3
M3	J+K+L	= SUM (J3:L3)
O3	I * M	= I3 * M3
P3	I * N	= I3 * N3
O5	O3+O4	= SUM (O3:O4)

We originally exported the estimate from RS Means online to Excel (Estimate Action). The exported Excel sheet has all the cost details. We took some of the cost information to create the one shown in Figure 2.2.

A Word of Caution

When copying a cell that contains a formula and then pasting it in a different location, the pasted equation adjusts automatically to the new location, in both row and column. For example, if the equation in cell H3, "= E3 * F3 * G3/12/27," is pasted into cell H6, it will read "= E6 * F6 * G6/12/27" (going three positions down); if it is pasted into cell N3, it will read "= K3 * L3 * M3/12/27" (going six positions to the right). If this was not the intention, the user can utilize some Excel tools or functions to reflect the specific intention. If any character in an equation representing a row or column is preceded by the symbol "$," its value will not be adjusted. For example "K$5" will adjust the column but will not adjust the row. "$K5" will adjust the row but will not adjust the column. "K5" will not adjust for row or column. The user can also utilize Excel's Special Paste feature that allows pasting formulas, values, formats, other attributes, or all. In all cases of copying and pasting, the user must ensure that the pasted materials reflect the true intention accurately. The paste function in Excel includes several options, such as pasting the values or format only.

The IF statement can be a useful function in spreadsheets. For example, when estimating footings, you may create a column with the title, Footing Formed? (Y or N). Let's say you are in cell E9. Assume you have the length, width, and depth in cells B9, C9, and D9, respectively. In the cell where you want the formwork amount, say F9, you can write the formula in the form of IF (condition, then, else). For example, in the statement: IF (E9 = "Y", (B9 + C9) * 2 * D9, 0), the computer will check the character in cell E9. If it is equal to the letter Y, then it returns the values of the operation (B9 + C9) * 2 * D9. If it equals anything other than Y, it returns the value 0.

Spreadsheet programs use many more mathematical, statistical, logical, and other functions. Review the Help section of the particular spreadsheet program you are using for detailed information on formulas.

Using RSMeans Online Estimating

RSMeans Online estimating contains all the information found in the *Building Construction Cost Data*. In addition to cost data, RSMeans online estimating contains several reporting and export options under the Estimate Action menu:*

1. Advanced Report
2. Email Report
3. Export to Excel
4. Report

The next steps demonstrate the solution of Example 1 using RSMeans online estimating:

1. Once RSMeans Online estimating is accessed, you will be prompted to the settings page (Welcome to RSMeans Online!). After choosing your preferences, click the "Save & Continue" button. You can uncheck the "Display Cost Data Preferences at start-up" box if you like to skip this page in your next login. You can still change your preferences later at any time.

2. Click on Manage Estimates on the top of the screen and then choose Create New Estimate. You will see a new screen with boxes for the new estimate information. Name the new estimate Example 1 and complete the information. For location, choose the United States, Florida, and then Miami. When finished, click Save & Continue.

---

*RS Means periodically updates the software. For this reason, some functions many differ slightly.

3. You will be prompted to the "Search Data" screen with the new estimate on the bottom of the right-hand side and the database on the top of it. Next time you log in, you will be prompted to this screen without any open estimate. You can go to Manage Estimates and double-click any existing estimate to open it or create a new estimate (Figure 2.4).

Figure 2.4

4. The new estimate will be open in the bottom part of the screen, part (1), but it is empty so far. Above the estimate, you will see the database. You can navigate in the database either by scrolling up and down in part (2), expanding divisions and subdivisions in part (3), or search in box (4). Note that you can use different search options: All, Any, Exact, Line, or Advanced. This includes search by description or number. Type item 033053400920 (no separation between numbers). You will see this item only on the right-hand side of the screen. If you typed part of the number, for example 03305340, you will see all the items under that section / group. Select the item by checking the box on its left side and then add it to the estimate by clicking the + sign in the top of the Estimate pane. The item appears in the estimate with 0 quantity. Enter the quantity 88.89 in the Qty box and hit the Enter key on your keyboard. The cost numbers change to reflect the quantity you entered.

5. Go to the second item, 03 30 53.40 3850. Since the second item is very close in location to the first one, scroll down until you see it, and then click it and add it to the estimate. Enter the quantity of the second item: 106.67. Now your spreadsheet contains both items. It should look similar to the example.

Also, some numbers (especially totals) may not appear entirely because the column is too narrow. You can expand the column by dragging the divider lines between columns sideways to reveal the complete numbers, just like in Excel. You may have to scroll to the right to see all the columns.

6. You can add items from the database to the estimate by checking the item in the database and then click the + button to add it to the estimate. You can add items using Add Favorite function that allows you to choose among items you already placed in your Favorites list (the blue star on the left of the item.) This function is useful for items used frequently. You can also use the "Insert custom line" function to add an item that you create for this estimate. If you want to remove an item from the estimate, check the item box on the far left of the line, and then click the Remove from Estimate button, which looks like a trash can.

7. You can choose among different options in the Estimate Action such as Advanced Report, Change Cost Data, Change Estimate Header, Email Report, Export to Excel, or Report. The user will have more flexibility when using the Advanced Report option; which is part of the RSMeans Online estimating program, or Excel; which is an external (third-party) program.

8. In order to calculate Total Cost Incl. O&P with values different from those used by RSMeans,[2] you can do so after you export the estimate to Microsoft Excel. Note that the functions (addition, multiplication, etc.) have transferred as well.

For this example, assume:

The total bare cost will be in columns N, O, P, Q for materials, labor, equipment, and total, respectively. (You will notice that some columns are hidden for better display in this book.)

- Go to the next unused column, AD.
- In cell AD4, give a title Ext. Total O&P 2.
- In cell AD5, insert the equation: $=N5*1.1+O5*1.60+P5*1.1$
- You will get the value \$92,084
- Copy this formula to AD6.
- In cell AD8, use the summation procedure for the cells above.

---

2. This will be explained later, but for the sake of simplicity, let's take it as bare cost + 10% for materials and equipment cost + 60% for labor.

In the Excel file, delete any columns you don't need in your estimate. Be careful not to delete columns containing data used for computations in other columns (Figure 2.5).

Quantity	LineNumber	Description	Crew	Daily Output	Labor Hours	Unit	Material	Labor	Equipment	Total	Ext. Mat.	Ext. Labor	Ext. Equip.	Ext. Total	Ext. Total O&P	Ext. Total O&P 2
88.89	033053400920	Structural concrete, in place, column (4000 psi), square, up to 2% reinforcing by area, 24" x 24", includes forms(4 uses), Grade 60 rebar, concrete (Portland cement Type I), placing and finishing	C14A	17.71	11.29	C.Y.	$ 376.68	$357.28	$45.40	$779.36	$33,483	$31,759	$4,036	$69,277	$89,941	$92,084
106.67	033053403850	Structural concrete, in place, spread footing (3000 psi), over 5 C.Y., includes forms(4 uses), Grade 60 rebar, concrete (Portland cement Type I), placing and finishing	C14C	75	1.49	C.Y.	$ 176.08	$ 44.98	$ 0.36	$221.42	$18,782	$4,798	$38	$23,619	$28,137	$28,380
										Total	$1,001	$52,266	$36,557	$4,074	$92,896	$118,078	$120,464

Figure 2.5

9. Under the Manage Estimates tab, you can create and name folders in order to organize your estimates. You can Move, Copy, Merge, or Delete estimates also.

Square Foot Estimator

The Square Foot Estimator tab contains square foot costs for different types of buildings with different options, mostly with a drop-down menu. This section reports approximate costs for various types of based on area, volume, or functional unit. More details will be discussed in Chapter 15.

Chapter 2 Exercises—Set A

For solutions to the exercises in Set A, see the link to the student companion website at www.wiley.com/go/constructionestim5e.

These exercises (Set A and Set B) focus on setting up electronic spreadsheets. Currently, Microsoft Excel is the most used spreadsheet software; however, you may use your favorite software. The equation formats might differ, depending on the software used. However, the results must be the same. Exercises using RSMeans CostWorks will be included in later chapters.

1. Set up a spreadsheet to calculate the amount of carpet we need for the following rooms:

Size of room	No. of rooms
10′×11′	13
10′×12′-4″	16
12′-6″×18′	8
15′-3″×16′	6
13′×21′	5

Add 5 percent waste factor.

2. Set up a spreadsheet to calculate the total cost of electricity someone paid over the course of a year if the cost for the months January through December are: $123.70, 110.65, 98.32, 95.20, 115.83, 141.44, 155.90, 162.18, 127.36, 97.79, 102.37, and 117.30. Also calculate the monthly average, least bill, and highest bill.

3. Set up a spreadsheet to calculate the volume of 40 square columns of concrete. Each column is 16″×16″ in section and 12′ long. Make the total in cubic yards.

4. Set up a spreadsheet to calculate the volume of 50 circular columns of concrete. Each column is 20″ in diameter and 12′ long. Make the total in cubic yards.

5. Suppose you have a random sample of people, with the sex, height, and weight of each person. Leave the names unsorted, and set up a spreadsheet to calculate the average height and weight for men and women. Use Data/Sort to separate the two sexes.

Observation	Sex	Height (ft)	Weight (lb)	Observation	Sex	Height (ft)	Weight (lb)
1	M	6.1	185	11	M	5.9	166
2	M	5.8	175	12	F	5.4	123
3	F	5.2	125	13	M	5.75	160
4	M	5.7	180	14	F	5.1	110
5	F	4.9	110	15	F	5	122
6	F	5.5	135	16	M	6.3	240
7	F	5.4	150	17	F	5.25	150
8	M	6.2	210	18	F	5.35	136
9	F	5.3	140	19	M	6.25	198
10	M	6	205	20	M	5.6	187

6. Set up a spreadsheet to calculate the total cost of clothing items you are buying from a department store. You bought one suit ($299), three pairs of pants (two @ $69 and one @ $59), four dress shirts (two @ $50 and two @ $40), and two T-shirts @ $20 each. There is a discount of 40 percent off suits, 35 percent off pants, and 25 percent off shirts (dress and Ts). There is a 7 percent sales tax on everything.

Chapter 2 Exercises—Set B

Solutions to Set B exercises are available to instructors only; see the link to the instructor's companion website at www.wiley.com/go/ constructionestim5e.

1. Set up a spreadsheet to calculate the final grades for students in a cost-estimating course based on:

   A. Homework: There are eight sets of homework, each worth 10 points.

   B. Computer project is worth 100 points.

   C. Two midterm exams are each worth 100 points.

   D. The final exam is worth 100 points.

   The final grade will be based on: 10% homework, 15% computer project, 17.5% for each midterm exam, and 40% for the final exam.

2. In Exercise 1, set up an IF statement to assign letter grades based on the scale:

   If total point grade ≥ 90%, then the letter grade = A.

   If total point grade ≥ 80% but < 90%, then the letter grade = B.

If total point grade ≥ 70% but < 80%, then the letter grade = C.

If total point grade ≥ 60% but < 70%, then the letter grade = D.

If total point grade < 60%, then the letter grade = F.

3. Go to a website that provides the current value of Dow Jones Industrial Average and record the value every 30 minutes, starting at 9 a.m. (on a workday) until 4 p.m. Record these readings in a spreadsheet along with the reading time, and then set up equations to calculate the minimum, maximum, and average values.

4. Set up a spreadsheet to calculate the volume of 48 rectangular columns of concrete. Each column is 20″ × 12″ in cross section and 12′ long. Make the total in cubic yards.

5. In the spreadsheet of Exercise 4, calculate the area of the formwork. The formwork needed for each column is equal to the surface area of all four sides.

6. In the spreadsheet of Exercise 4, calculate the weight of the reinforcement steel, assuming that each column has eight bars of #10. Assume the length of the bars to be 14′ each (2′ for the splices). No. 10 reinforcement bar weighs 4.303 lb/LF.

Chapter 3

Cost Estimating: An Introduction

Introduction   Cost estimating is one component of cost management, which is a major component in construction project management. Organizations and companies have different approaches to how cost estimating and management fit into their project management model, but there is a consensus on the fact that cost management is one of the most important components of project management.

As the name implies, cost estimating aims at *predicting future* expenditures for a certain project or scope. This prediction will take several iterations, as explained later in this chapter. Like a camera lens, the more you focus on the target, the clearer the picture gets. One difference, though: No matter how accurate the cost estimate is, it will never be 100 percent accurate in the sense of exactly matching later actual expenditures. This leads us to the other component of cost management—namely, cost control, which can be defined simply as comparing the estimate with reality, finding variances or deviations and their causes, and taking corrective actions to bring the project back within budget.

Cost estimates can be prepared by an owner, a design consultant, a professional construction management consultant, a contractor, or other person.

Cost management is related to, and integrated with, many facets of project management, such as scope management, time management (planning and scheduling), project controls, procurement management, quality management, risk management, accounting, budgeting, financing, value engineering, logistics, safety, and other functions. Cost management is extremely important for the success and continuing operation of construction companies.

Definitions   In a simple definition, *cost* is the total expenditures in dollars (or any local currency) incurred in the completion of a project. *Cost estimating* is defined as the predictive process used to quantify, cost, and price the resources required by the scope of an investment option, activity, or project. Cost estimating is a process used to predict uncertain future costs. In this regard, a goal of cost estimating is to minimize the uncertainty of the estimate given the level and quality of scope definition. The outcome of cost estimating ideally includes both an expected cost and a probabilistic cost distribution. As a predictive process, historical reference cost data (where applicable) improve the reliability of cost estimating. Cost estimating, by providing the basis for budgets, also shares a goal with cost control of maximizing the probability of the actual cost outcome being the same as predicted.[1] In a simpler definition, cost estimating is the process used to quantify, cost, and price the resources required by the scope of an investment option or a project.

An important note on actual cost that even some contractors overlook: The *real* actual cost may not be completely known until the end of the warranty period, not the physical completion of the project.

Cost estimates are determined using experience; calculating and forecasting the future cost of resources, methods, and management within a scheduled time frame. Detailed estimates require quantitative analysis of the work required by the design documents.

Quantity estimating, by contrast, is the process of measuring the work items of the project in the form of a series of quantified work items, such as item count, linear feet, area, or volume. This includes work items to be performed (quantities estimated) and work items already performed (quantities calculated/determined). In other words, quantity estimating means calculating the quantities of materials required for the project; before, during, and after the execution, including change (variation) orders and what-if scenarios. In the United Kingdom, the quantity surveyor (usually abbreviated as QS) is a lot more than a quantity estimator. The QS is even more than an estimator, probably more like a cost manager. There are colleges in the United Kingdom and other British Commonwealth countries where people obtain degrees in quantity surveying, but curricula usually include, in addition to quantity surveying and estimating, courses in contract management, planning and scheduling, project management, law, and other related construction fields.

---

1. AACE International. *Recommended Practice No. 10S-90, Cost Engineering Terminology* (Morgantown, WV: Author, November 14, 2014). http://www.aacei.org/toc/toc_10S-90.pdf.

In US terminology, the terms *quantity estimating*, *quantity surveying*, and *quantity takeoff* have almost the same meaning. They all mean the process of collecting the information (quantity and type) on all materials and equipment included in a project using the design drawings and specifications plus any additional contractual documents. This information is the foundation the cost estimator depends on for estimating the cost of the project. The term *takeoff* is used both as a verb and noun. Quantity estimating or takeoff requires many skills, especially mathematical ones: arithmetic, geometric, trigonometric, and algebraic. It also requires sharp eyes and good observation and coordination skills. For simplicity, the author will be using the term *quantity estimating* in this book but may alternate among the three terms.

Cost estimating is a lot more than quantity estimating. On one hand, quantity surveyors/estimators have to read and interpret contract documents—namely, drawings (blueprints) and specifications. From these documents, they calculate quantities, such as the total cubic yards/meters of excavation, cubic yards/meters of concrete, square feet/meters of concrete formwork, number of bricks, lineal feet/meters of electric wires, and so on. They need to be accurate and specific in such quantities. For example, it is important to specify type of excavation (classification of soil, conditions, moisture content), the type of concrete (strength, additives, special applications) and the type of concrete member, types of bricks, types of electrical wire, and others. In order to make sure to account for every item in the project with the correct description, quantity estimators need to do other tasks such as communicating with the A/E (architect/engineer) and vendors.[2] The cost estimator, on the other hand, deals with quantity estimating, as already described, in addition to estimating the prices of labor, equipment, and indirect expenses (discussed in detail later). This estimating may be challenging and critical, taking in consideration the variation of labor productivity, anticipated escalation and inflation of prices, other risks that impact the cost, and many other factor and issues. In addition, cost estimators are usually tasked with related work, such as negotiating with client, subs, vendors, and others; preparing bids and leading the bidding process; and providing input and interacting with others team members of the project management.

One major difference between quantity and cost estimators is the impact of change in time or location. This change has little impact on the quantity estimators' job but has a major impact on the cost estimators' job. Cost

---

2. Although A/E is an abbreviation for architect/engineer, it is a designation for the designer of the project. It is a term used generally in the building construction industry.

estimators have to be up to date on material and equipment prices, labor wages, interest rate, insurance and bonds, and other cost items. It is very common for construction companies to see the cost estimating team headed by a cost estimating manager with quantity surveyors on the team.

Schedule of values is a listing of elements, systems, items, or other subdivisions of the work, establishing a value for each, the total of which equals the contract sum. The schedule of values is used for establishing the cash flow of a project and serves as the basis for payments requests.[3] It is also called *project cost breakdown* or simply *cost breakdown*, or *bill of quantities* (BOQ) in the United Kingdom, and serves as the basis for payments requests and may help in determining the cost of change orders.

Cost and value: When the term *cost* is used, it usually refers to the amount of money, cash expended, or liability incurred, spent in consideration of goods and/or services received at the time this consideration occurs. In contrast, *value* (or economic value) is a measure of the benefit provided to the owner by the good or service over its life. Some define value as

$$\text{Value} = \frac{\text{Perceived benefits received}}{\text{Perceived price paid}}$$

Realizing the difference between cost and value is important to the owner so he or she can make the best decision based on life cycle value and not just the up-front cost. For example, the value of a shopping center project is higher to the owner if completed in October rather than January. This is why a clear explanation of the difference by the design professional is important.

Types and Purposes of Estimates

Cost estimates have been classified into several types using different criteria. Estimates have different levels of accuracy depending on several factors, such as end use (purpose of the estimate), level of effort used in preparing the estimate, and the availability and degree of completion of the design documents. The AACE International produced a *cost matrix* classifying estimates into five classes according to four criteria: The primary characteristic is maturity level of project definition deliverables (expressed as percentage of complete definition) and the secondary characteristics are end usage (purpose of estimate), methodology (estimating method), and expected accuracy range (see Figure 3.1).[4]

---

3. *RSMeans Illustrated Construction Dictionary,* 4th ed., unabridged (Kingston, MA: RSMeans, 2009).
4. AACE International, *Recommended Practice No. 56R-08, Cost Estimate Classification System—As Applied for the Building and General Construction Industries,* TCM Framework: 7.3—Cost Estimating and Budgeting, revised March 1, 2016.

18R-97: Cost Estimate Classification System – As Applied in Engineering, Procurement, and   3 of 15
Construction for the Process Industries, revised March 6, 2019

March 1, 2016

ESTIMATE CLASS	Primary Characteristic	Secondary Characteristic		
	MATURITY LEVEL OF PROJECT DEFINITION DELIVERABLES Expressed as % of complete definition	END USAGE Typical purpose of estimate	METHODOLOGY Typical estimating method	EXPECTED ACCURACY RANGE Typical variation in low and high ranges
Class 5	0% to 2%	Concept screening	Capacity factored, parametric models, judgment, or analogy	L: −20% to −50% H: +30% to +100%
Class 4	1% to 15%	Study or feasibility	Equipment factored or parametric models	L: −15% to −30% H: +20% to +50%
Class 3	10% to 40%	Budget authorization or control	Semi-detailed unit costs with assembly level line items	L: −10% to −20% H: +10% to +30%
Class 2	30% to 75%	Control or bid/tender	Detailed unit cost with forced detailed take-off	L: −5% to −15% H: +5% to +20%
Class 1	65% to 100%	Check estimate or bid/tender	Detailed unit cost with detailed take-off	L: −3% to −10% H: +3% to +15%

Figure 3.1 AACE International Cost Estimate Classification Matrix for Building and General Construction Industries. Reprinted with the permission of AACE International, 209 Prairie Ave., Suite 100, Morgantown, WV 25601 USA. Phone 800-858-COST/304-296-8444. Fax: 304-291-5728. Internet: http://www.aacei.org. E-mail: info@aacei.org Copyright © 2007 by AACE International; all rights reserved.

In addition to the degree of project definition, estimate accuracy is also driven by other systemic risks, such as:

- Complexity of the project.
- Quality of reference cost estimating data.
- Quality of assumptions used in preparing the estimate.
- Experience and skill level of the estimator.
- Estimating techniques employed.
- Time and level of effort budgeted to prepare the estimate.

Systemic risks such as these are often the primary driver of accuracy; however, project-specific risks (e.g., risk events) also drive the accuracy range.[5]

These four characteristics in the cost matrix, in addition to the preparation effort, are interrelated and/or interdependent.[6] Some of them are controlled by the estimator while others are not. For example, the estimator does not usually control the *degree of project definition*. Design

---

5. Ibid.

6. Preparation effort was the fifth characteristic in the matrix before it was modified in 2011.

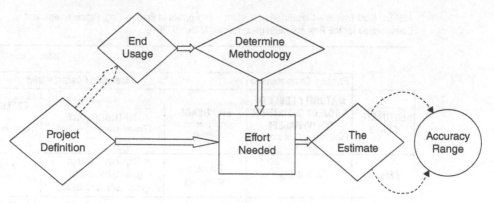

Figure 3.2   Factors Influencing Type of Estimate and Accuracy Range

information comes from the owner or design consultant and might not be at a level that allows the estimator to create or obtain a detailed estimate. The estimator decides the *end usage* for the estimate, which controls—or has great impact on—the methodology, which, in turn, requires a certain *degree of project definition* and *level of effort*. This effort results in an estimate with an *expected accuracy range* (see Figure 3.2).

There is a direct but nonlinear proportionality between effort (time and money) spent on preparing the estimate and its accuracy, as shown in Figure 3.3. This relationship, of course, is within the limitation of the availability and degree of definition of the project's design documents. Even with high degree of project definition, the contractor may need only an approximate estimate using the square-foot estimating methodology,

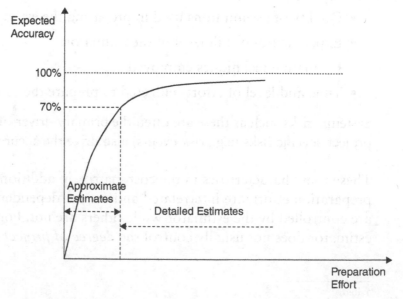

Figure 3.3   Relationship between Preparation Effort and Expected Accuracy

which needs little preparation effort, resulting in an estimate with a low range of accuracy. However, an estimator cannot do a detailed estimate with a low degree of project definition.

If we classify estimates simply as approximate and detailed, we will find a fundamental difference in the methodology between the two types. Approximate estimates are usually prepared in very short time by multiplying the area (or number of functional units) in the proposed project by a predetermined unit price and then applying needed adjustments. This predetermined unit price may have been obtained from one's own previous experience in similar projects or from third-party information, such as the RSMeans data books. Comparatively, detailed estimates require quantity takeoffs of all work items in the proposed project and their specifications. This process cannot occur without the availability of the detailed design, and it takes a high level of effort.

These two types of estimates (approximate and detailed) are not exactly discrete; there are multiple degrees of accuracy between them. Let's use this analogy: There is a 100-piece picture puzzle, and you are trying to put it together. You don't have all 100 pieces so you are using the available pieces and complementing them with your imagination (assumptions) of what the picture would look like. The picture you are obtaining is fuzzy (approximate, not definitive). Now assume you are receiving more pieces of the puzzle, a few at a time. Every batch of pieces you receive will be used in the puzzle, replacing the imagined ones and, thus, making the picture clearer. Project design goes through stages, moving from an idea to conceptual design, schematic design, detailed design (through several stages called design development, DD%, culminating in final design), and finally bidding documents. It is very common for an estimator to do multiple and successive estimates for the same project at different stages, where every stage has more information on the design, resulting in an increase to the accuracy of each new estimate.

Approximate estimates usually have an accuracy range of about –30 percent to +50 percent (i.e., 30 percent underbid to 50 percent overbid). Sometimes approximate estimates are called order of magnitude, preliminary, back-of-the-envelope, or summary estimates. Conceptual estimates are also a type of approximate estimates. Approximate estimates are used to:

- *Conduct feasibility studies.* An owner may do an approximate estimate to know the ballpark figure of the project's cost. This may help in the decision of *go/no go* for the project. An investor may prepare what is known as a pro forma.[7]

- *Compare alternative projects.* An owner/investor might want to compare several projects to choose the most profitable one. At this stage, only approximate estimates are needed.

- *Compare alternative designs.* After an owner has chosen a certain project, he or she may consider several design options, such as reinforced concrete versus structural steel, a concrete dome versus truss, and so on.

- *Make initial financial arrangements/budget.* An owner or a contractor may prepare an approximate estimate to determine whether the project is within its financial capability. Also, an approximate estimate can be used to make financial arrangements, such as applying for a loan or recruiting investors.

- *Provide a guaranteed maximum price (GMP) in cost-plus-fee contracts.* A contractor may take on a project using a cost-plus-fee contract (explained later in this chapter) but be required to provide the owner with an estimate of the maximum possible cost to the project.

Detailed estimates usually have with an accuracy range of about –5 percent to +15 percent. Detailed estimates are also called final or definitive estimates. They are used to do the following:

- *Bid on a project.* A contractor needs a detailed estimate to bid on a project, especially lump-sum bids.

- *Project control.* A detailed estimate is needed for cost control. Such an estimate is usually adopted as the baseline budget, then actual spending is compared to it.[8]

Some estimators and authors define a category called *budget estimates* with a degree of accuracy higher than approximate but lower than detailed estimates, usually around –10 percent to +20 percent.

In a quick survey of several references on types of cost estimating, we find a variety of types based on one of the criteria mentioned earlier: degree of project definition, end usage, methodology, and expected

---

7. *Pro forma* (written also as proforma) is a projection or estimate of the cost/benefit ratio or rate of return on an investment. Project owners and investors usually use a pro forma to help them make a decision on whether to carry out the project or not. Typically, a pro forma includes all expected costs (initial, recurring, and occasional) and expected revenues throughout the life cycle of the project, then it calculates the rate of return.

8. A baseline budget is the project's original approved budget, including any approved changes, which is used to compare actual spending to. Usually, the initial (before construction) detailed estimate forms the basis of the baseline budget, but it may get modified later on during the contracting phase or due to change orders.

accuracy range. According to the American Society of Professional Estimators, "The levels of the estimate correspond to the typical design process and are considered standard within the industry. These levels are as follows: Order of Magnitude, Schematic/Conceptual Design, Design Development, Construction Document, Bid."[9] But as we notice, the first-level order of magnitude indicates expected accuracy range, the next three levels refer to degree of project definition, and the last level indicates an end usage.

RSMeans recognizes four major types of estimates:

1. Order of magnitude or general SF estimate: Used for general project budget forecast.
2. Square foot estimate: Usually needed for financing.
3. Assemblies estimate: Used during design development to evaluate system costs against the project budget.
4. Unit price estimate: When detailed cost breakdown is needed, such as for bids, contract documents, or change orders.

Assembly and model estimating are similar in concept. They both combine related items that make up an assembly, which is part of the project, or a model, which is the entire project. Model estimates include both items and assemblies. Such methods of estimating became more powerful and popular with the use of computers because users can usually choose among several alternatives (in a drop-down menu) to customize the assembly or the model to the specific need.

Square-foot and functional unit estimating are approximate methods based on comparison to the unit cost of similar previous projects. The reference number is obtained by dividing the total cost of the known project (or similar projects) by the total area (square feet or square meters) to produce a unit price. This unit price is multiplied by the area (in square feet or square meters) of the proposed project. Adjustments then need to be applied for factors such as location, level of finish, inflation, and so on. It is also customary to use a functional unit other than the square feet, such as bed (for hospitals, nursing homes, and dormitories), key (for hotel room or suit), pupil (for schools), car space (for parking garage), or apartment (for large apartment building/complex). Caution must be exercised here because the variation may be large between the base being compared to and the proposed project.

---

9. American Society of Professional Estimators, *Standard Estimating Practice*, 9th ed. (Vista, CA: BNi Publications, 2014).

Parametric estimating is another concept:

> [It]uses a statistical relationship between relevant historical data and other variables (e.g., square footage in construction) to calculate a cost estimate for project work. This technique can produce higher levels of accuracy depending upon the sophistication and underlying data built into the model. Parametric cost estimates can be applied to a total project or to segments of a project, in conjunction with other estimating methods.[10]

A bottom-up estimate is another type of estimates or, rather, estimating methodology. The cost of individual work packages or activities is estimated to the greatest level of specified detail.[11] Bottom-up estimates often utilize the work breakdown structure (WBS), where the project is broken down in a progressive manner; every level is more detailed than the previous one (see example in the next section, "Format of the Estimate"). The detailed cost is based on the lowest (most detailed) level and then summarized or *rolled up* to higher levels for subsequent reporting and tracking purposes. The cost and accuracy of bottom-up cost estimating is typically influenced by the size and complexity of work packages.

Rules for the "Good Cost Estimator"

1. The good estimator must be able to come up with an approximate cost (ballpark figure) for a project, with primitive or no tools available in a reasonable amount of time.

2. While the price of one item may be too high or too low, the overall estimate should be pretty accurate.

3. The good estimator must have the intuition to make mental judgment on numbers and tell if the numbers are too low or too high.

Format of the Estimate

In the past, estimators used to organize their estimates based on their own preference or their company's policy. Later on, the Construction Specifications Institute (CSI) came up with an organization called MasterFormat 1995 that included 16 divisions. This format was popular and became the standard for most estimates in the construction industry. However, many items did not clearly fit any of these divisions, such as

---

10. *A Guide to the Project Management Body of Knowledge (PMBOK® Guide)*, 6th ed. (Newton Square, PA: Project Management Institute, Inc., 2017).

11. Ibid.

fiber optic cables and security systems. Besides, with the advancement of science and technology, many new items emerged that also did not fit any of the existing divisions. In 2004, the CSI changed and expanded its MasterFormat to include 50 divisions, many of them unnamed but reserved for future use. This MasterFormat had minor revisions since 2010, with more sections being added and the titles of some sections being changed for clarification. Both MasterFormats (original 1995 and current 2016) can be seen in Appendix C.

Work Breakdown Structure

Work breakdown structure (WBS) is defined as a task-oriented detailed breakdown of activities that organizes, defines, and graphically displays the total work to be accomplished in order to achieve the final objectives of a project. The WBS breaks down the project into progressively detailed levels. Each descending level represents an increasingly detailed definition of a project component. In project scheduling, the components at the lowest WBS level are often used as activities to build the project schedule.[12]

There is no universal WBS format, but the new CSI MasterFormat is compatible with WBS. Table 3.1 is an example of WBS out of the CSI MasterFormat 2016 with five levels.

Keep in mind that there is an even higher WBS level than level 1, which is level 0, and there is only one item in the estimate at that level—the entire

Table 3.1   Partial Breakdown Taken from the CSI MasterFormat 2016
For more details, go to http://www.csinet.org/

Title	Level
03 00 00 Concrete	1
03 10 00 Concrete Forming and Accessories	2
03 11 00 Concrete Forming	3
03 11 13 Structural Cast-in-Place Concrete Forming	4
03 11 13.13 Concrete Slip Forming	5
03 11 13.16 Concrete Shoring	5
03 11 13.19 Falsework	5
03 11 16 Architectural Cast-in Place Concrete Forming	4
03 11 16.13 Concrete Form Liners	5
03 11 19 Insulating Concrete Forming	4
03 11 23 Permanent Stair Forming	4

---

12. Saleh Mubarak, *Construction Project Scheduling and Control*, 3rd ed. (Hoboken, NJ: John Wiley & Sons, 2015), p. 49.

project. Estimated items will be at the lowest level (5 in this example); then they can be rolled up to any higher level for the purposes of control and reporting.

Sources of Cost Estimating Databases

The best database for a cost estimating professional or construction company is its own, if such database exists and provided it is well organized and regularly updated. In many situations, the estimator may not know the price of an item because it is a new item, it is a known item but in a different location or country, the market has changed since the estimator last priced it, or for any other reason. The estimator can use commercial third-party sources, such as the RSMeans data books and Richardson's in the United States, or Davis Langdon's Spon's database in the United Kingdom and NODOC—Nomitech Database of Cost for the oil and gas industry.

Software vendors also create their own databases but allow users to either build their own database or tie in to a third-party commercial database, such as RSMeans. With the advancement of technology, software users can update databases (prices, productivities, and other data) online.

Building your own database can be a difficult and time-consuming task. It will most likely require contributions from several specialists. Maintaining the database is another monumental task. It has to be kept up to date in terms of item availability, prices, productivities, and other information. Items may be taken out of the database while new items are added. Databases can be resident on the hard drive of a personal computer in the simplest form, or they can be on the server and may be accessed online with proper security procedures. Security and backup of databases are extremely important.

Evolution of Cost Estimating Tools

Construction cost estimating is as old as construction itself, and its methods and tools have evolved through history. Takeoff methods were manual with basic, and simple tools until the past few decades. In the very old times, construction cost estimating was mostly practiced by mental "guestimation" by the builder based on previous experience. Later, when design professionals started producing design documents, contractors and estimators used these documents for estimating, employing simple tools such as rulers and pencil and paper. The invention and advancement of computers provided a significant boost not only to cost estimating but also to all facets of the building construction activities starting from the design drawings.

Takeoff began, as mentioned earlier, by using intuition and experience of the contractor in calculating quantities, along with mental and manual mathematical computations. Simple tools were utilized later, such as rulers, scales, and pencils, along with paper spreadsheets and manual and slide rule[13] computations. Manual roller pens that calculate the length of a line in the drawings were invented using the same principle as the odometer. In the early 1970s, hand calculators were invented and have continuously gotten better and more sophisticated, calculating areas and volumes and doing unit conversions. Digital rollers pens then were introduced. Soon after came digitizers. These digitizers were very popular in the 1990s and 2000s. They enabled estimators to convert paper dimensions into digital formats with the simple use of a handheld stylus. The estimator simply laid down the drawings on the digitizer and picked up lengths and areas, including odd and difficult shapes, by simply touching or dragging a stylus. These digitizers were linked to computer software programs that took the quantities directly to the estimate and showed the area or line picked up by the digitizer directly on the computer screen, minimizing the possibility of dropping or duplicating an item. Estimators could choose between several sizes and types of digitizers, including hard and flexible. They are still in use, but their popularity has been declining steadily with the increased use of electronic takeoffs directly from CAD drawings and building information modeling (BIM).

At the same time, computer hardware and software have been advancing by leaps and bounds. Computer-aided design (or drafting), abbreviated as CAD, has revolutionized the design process and cost estimating along with it. It started with two dimensions (2-D) then moved to three dimensions (3-D). New programs allow the direct takeoff from CAD drawings to the estimate. Recently, BIM has become the new trend in bringing several elements of the design and construction to one platform in order to simulate the real activities. BIM allows the estimating process to become more efficient and detect clashes before they even happen. BIM started with 3-D design but added more "dimensions," starting with time and then adding other criteria, such as cost, specifications, and others. "The BIM platform serves as a central location for both the construction and design teams to streamline their activity and provide more insightful and accurate reports."[14]

---

13. Slide rules are mechanical analog computers that were invented in the seventeenth century and were successively improved and updated. They were popular for computations until the middle of the twentieth century, when pocket calculators were invented. By the mid-1970s, slide rules had become obsolete.
14. Hendrick Degenaar, "The Evolution of Cost Estimating," *Construction Today* (online), www .construction-today.com/index.php/sections/columns/1059-the-evolution-of-cost-estimating.

The next generation of BIM can access design and construction models through a broad range of cloud-based systems that provide mobility, accessibility, and virtually infinite computing power. Multidiscipline design and construction teams can improve project outcomes by moving computation-intensive tasks to the cloud, enabling more rapid simulation and visualization and optimized collaboration with access to intelligent, data-rich models. This will have a great impact on construction cost estimating.[15]

Computer estimating has come a long way, too. It started with computer spreadsheets, such as Lotus 123, Quattro-Pro, and Excel. Spreadsheets were used first to store item pricing and other information (mostly as templates). Later on, databases were utilized and connected to estimates, which led to professional cost estimating software, combining spreadsheets and databases in one program. At first, information in the databases had to be updated manually either by having the estimator change prices manually or by copying updated files from a floppy disk. Updates later took different shapes, such as getting direct updates from the vendor via telephone lines and modems and later on through the Internet.

Estimating programs also got more sophisticated in several aspects, including the features and functionality of the software. Also, assembly and model estimating was introduced, which saved time and gave flexibility to the estimator. For example, when the contractor does repetitive construction, he or she can have a detailed estimate for a building as an assembly. When estimating a similar building or assembly, the model can be customized to fit the specifications of the new building, and prices get updated to match those currently in the database. The estimator can sit with the client discussing the estimate and exploring many what-if scenarios (such as different types of building façade, flooring, wall covering, doors and windows, plumbing fixture brands, etc.) with the cost estimate changing instantly to reflect the project with the chosen options.

There is no question that the technological advancements in quantity takeoff tools and in cost estimating software and databases have contributed to improving the accuracy of cost estimates as well as to shortening the time required for preparing these estimates.

Project Costs to an Owner

Every time the term *cost of a construction project* is mentioned, many people think of just the cost of the design and construction (building) of

---

15. Ibid.

that project and the land, if needed. In reality, there is a lot more to it from the owner's perspective than just these two costs. A list of 13 costs that are usually associated with any project follows.

1. *Lost opportunity cost.* This represents the amount of profit that would have been earned if the money was invested differently.

2. *Land, right-of-way (access), easements.* In addition to the cost of the land, the owner may have additional related expenses, such as right-of-way or easement through a neighboring adjacent land. Another point to consider is that in many situations, the law defines the *purchase* of the land as a long-term lease (usually for 99 years).

3. *Financing (cost of borrowing money).* The owner may pay financing costs for the project in more than one way. There are owner's purchases and services, including construction and land, that usually require financing. The traditional way of contractor spending and owner reimbursement usually results in additional financing cost to the contractor, which is passed to the owner (see Appendix B: Contractor's Cash Flow).

4. *Design.* This includes the architectural, civil, structural, mechanical, electrical, and other services and may also include government permitting fees. These fees range usually between 5 and 12 percent of the estimated project construction cost, depending on several factors.[16]

5. *Permits, licenses, and fees.* Such fees may be initially paid by the owner, the design consultant, or the contractor; but eventually they come out of the owner's pocket. Developers usually pay impact fees to local governments as compensation for the impact of adding a new development on the existing infrastructure and services the government provides.

6. *Construction.* This includes all expenses associated with the construction process (see "Direct versus Indirect Expenses" later in this chapter).

7. *Project/construction management services.* This is yet another service the owner may need to ensure that the project is being executed according to the contractual agreement and that the owner's interest is protected. The term *project management services* is wider and more comprehensive than *construction management services*, as it starts earlier—perhaps as early as the project is initiated. Sometimes the term *supervision* is used in lieu of *construction management*.

---

16. In non-Western countries, this percentage may be significantly less.

8. *Furniture and equipment.* This cost has to be considered by the owner or someone else (possibly the tenant for residential and commercial projects). Depending on the style, quality, level, and other factors, this can be a significant expense.

9. *Start-up.* In many industrial projects, the owner requires start-up, which means that the contractor delivers the project to the owner in a working condition. It might include training personnel also.

10. *Warranty.* Most building codes require builders to warrant their projects for a minimum period of time. Some builders offer a longer or more comprehensive warranty as an attraction or confidence booster to customers. The owner may also require an extended warranty, but, of course, the longer this period is, the more expensive it will be, even though this cost might not be conspicuous to the owner.

11. *Operating and maintenance.* This may be a major expense to the owner. It includes operating and maintenance not only to the structure itself but also to the equipment and main systems affiliated with the building.

12. *Rehabilitation/upgrading.* In industries that change rapidly with technology, this can be a major expense. For example, if an office building has personal computers, they will need an upgrade on average every three years. In many cases, the infrastructure of the computers or other equipment may also need upgrading/updating.

13. *Disposal.* Disposal can be a major headache and expense, especially if the project contains toxic chemicals or radioactive materials. Disposal might also include demolition in a crowded downtown area, which is another major expense and liability. Hauling and dumping fees may also be substantial.

The total cost that includes these 13 items constitutes the *life-cycle cost* for a project. Professionals and experts sometimes conduct value engineering (VE) studies in order to optimize the project's total life-cycle cost.[17] This VE study may result in a slight increase of the up-front cost in exchange of bigger future savings and/or better performance. Also, BIM started to be utilized to simulate the life cycle (design, construction, operation) of the facility to optimize cost or other criteria.

---

17. Value engineering is a science that studies the relative value of various materials and construction techniques. It considers the initial cost of construction, coupled with the estimated cost of maintenance, energy use, life expectancy, and replacement cost.

To Bid or Not to Bid?

Detailed estimates are expensive to prepare. An estimating team may dedicate its time for several days to prepare a bid for a project that might later be won by a competitor. Contractors have a difficult decision to bid or not to bid a project. They assess the potential risk and profit associated with the considered project and then make a decision based on eight key factors:

1. *Type of project/construction involved compared to the contractor's area of expertise.* The contractor may, of course, subcontract any portion of the job that is outside its area of expertise (or practice) or that it feels can be performed more efficiently by a specialty subcontractor.

2. *Location of the project.* Contractors usually like to work on jobs within their region or where they can provide efficient services. Taking on a job far away from headquarters or regional offices may create logistical problems and result in higher cost, in addition to the possibility of the contractor being unfamiliar with that specific local market (vendors, suppliers, subcontractors, labor).

3. *Size of the project.* The contractor cannot take on a project beyond its bonding capacity and/or available cash or line of credit (borrowing capacity) unless it teams up with someone else to reach the required capacity.

4. *Owner and architect/engineer (A/E).* Reputation and previous experience with a particular owner and/or an A/E may influence the contractor's decision to bid or not to bid, or it might be reflected on the contractor's price tag. Contractors tend to lower their price (mainly fees) when there is a favorable relationship with the owner or A/E, and vice versa. Sometimes a contractor lowers its fees with a particular owner in hope of establishing a relationship for future work with this client.

5. *Amount of work currently under construction.* When economic conditions are favorable, contractors are usually busy, and the demand exceeds supply. In this case, contractors are less likely to bid on projects that are not their favorite types.

6. *Equipment, qualified personnel, and subcontractors required compared to those available to the contractor.* This is a matter of both feasibility and economy to the contractor. Theoretically, the contractor can acquire any item, but the cost has to be acceptable to keep the overall costs reasonable.

7. *Future strategic plans.* The project may be in a geographic area that is outside the contractor's base or even outside his or her area of expertise/preference, but the contractor may choose to bid on the project to expand business and fulfill the organization's strategic objectives.

8. *Other bidders (competitors).* Previous experience gives the contractor a feeling (a hunch) of his or her chances in winning the project against competitors. Knowing the names of competitors may encourage or discourage contractors from bidding.

All the preceding eight factors have to do with the risk the contractor is taking or willing to take in the project. Construction is a risky business, and there is no such thing as risk-free job. In general, profit is proportional to risk taken. However, the contractor must balance all factors in making the decision to bid or not to bid with the expected return and cost of bidding the project.

Sometimes it is not the contractor's choice to bid or not to bid. Owners' conditions and stipulations, such as bonding capacity, minimum experience in certain type of work, or security clearance, might disqualify some contractors. Private owners have more leeway in including or excluding certain contractors or even negotiating contracts with a certain contractor without bidding. This might happen for government projects only in legally justified cases, such as in national security and national defense contracts.

Types of Contract Award Methods

Construction contracts may be awarded from owners to contractors using one of these three methods:

1. *Competitive bidding*—when the owner publicly advertises the project, and every qualified contractor has an opportunity to bid. Owners qualify bidders before or after the bid opening (this is called pre- or post-qualification). In most competitive bidding events, owners require a minimum number of qualified bidders (in most cases, three).[18] If the number of bidders does not reach the minimum, the owner may cancel the bidding, extend the invitation to bidding, or negotiate the bid with the one or two bidders, if the regulations allow such practice.

---

18. Sometimes only one or two contractors (less than the minimum number required) qualify as bidders on a project. The bidding contractor(s) might ask another contractor (who did not intend to bid) to submit a complementary or courtesy bid. This is not a serious bid but a deliberately inflated one, intended to help other bid(s) meet the requirements. This practice is not ethical and may be illegal as well.

2. *Selective bidding*—when the owner invites a group of selected contractors to bid on a project. This usually happens in private projects but can also happen in government projects only if there is a legal justification, such as in national security or defense projects. Note that when this practice happens in the private sector, the owner can subjectively pick those contractors to be invited and may not necessarily award the project to the lowest bidder. The owner may award it to the bidder who succeeds to convince the owner that it will do the best value for the money.[19]

3. *Direct award*—when an owner chooses a particular contractor to do the project. This process may, and usually does, involve some negotiation between the two parties until they both agree on the contract amount as well as other contract terms. The process may not necessarily end up in an award. The two parties may fail to come to an agreement, in which case the owner may start negotiating with a different contractor. Again, this method is used in the private sector but can also happen in government projects only if there is a legal justification for such practice.

Many public agencies also have a policy of annual or biannual unit-price contracts with a group of officially approved contractors. This policy starts with the agency requesting unit price bids through an open invitation on certain items, such as excavation, drainage, paving, resurfacing, or sidewalks. Contractors who fulfill the bidding requirements and whose unit prices fall within the acceptable limits by the agency sign a one- or two-year contract with the agency for potential work during that period, with the unit prices specified in their contracts. When the agency has a need for work of a certain type with a total cost within a certain limit, it can award that job it to a contractor on the list based on the contracted unit prices. If that project has a total cost that exceeds the legal limit of this type of contract, the agency will be obligated to go through the normal bidding process (competitive bidding, as mentioned earlier). Also, public agencies usually award projects on a rotational basis, with each contractor having an annual quota for its total unit-price contracts for that year.

When Time Matters

In certain projects when time matters, such as in highway projects, the government agency likes to minimize the period of the highway being shut

---

19. This point underscores the important of good communication. The selective bidding process usually culminates, just before the award, with a presentation for each contractor (separately, of course). In this presentation, the contractor tries to answer the owner's question: Why should I hire you and not the others? In tight competitions, good communications skills may be tipping factor.

down or to encourage the contractor to work during off-peak construction hours for the convenience of the public. The owner or government agency may resort to traditional or other more creative contracting methods to pressure the contractor to complete the project by the completion date stipulated in the contract or expedite it as much as possible. The traditional method adds a clause in the contract, imposing penalty (called liquidated damages, or LDs for short) on the contractor for every day of delay beyond the contractually specified project duration. This penalty can be fixed (e.g., $5,000/day) or progressive, such as $2,000/day for the first week, $2,500/day for the second week, and $3,000 for every day after two weeks. The penalty may be coupled with a bonus for every day of completion earlier than the contractual completion date.[20]

Usually, but not always, there is a cap on the LDs and/or the bonus. LDs by nature are not supposed to be punitive to the contractor. Instead, LDs should represent the owner's actual/potential losses as a result of the delay. In public projects, such as highways, bridges, and government building, it is extremely difficult to quantify such losses so public agencies estimate LDs based on the importance of the project to the public and their inconvenience as a result of inability to use the facility or reduction in its capacity. There are always legal challenges to the amounts of these LDs.

Nontraditional and more creative methods that encourage the contractor to expedite the project and / or avoid work during high traffic hours include A + B and lane renting. The first one is a bidding/contracting method that factors time plus cost to determine the low bid. Under the A + B method, the owner, usually a public agency, sets a hypothetical cost per day of the project. Each bidder submits a bid with two components:

1. A: dollar amount for contract items (i.e., cost estimate).
2. B: days required to complete the project.

Bid days are multiplied by the set "cost per day" and added to the A component to obtain the total bid.

$$A + (B \times \text{Cost per day}) = \text{Total bid}$$

This is also called cost plus time or biparameter bidding/contracting method.

For example: There are three bidders for a project. The owner sets a value of $10,000/day of construction.

---

20. For more details and case studies, see Mubarak, *Construction Project Scheduling and Control* (3rd ed.), chapter 3, "Construction Delay and Other Claims," pp. 343–366.

Table 3.2 Example of Bid Analysis Using the A + B Contracting Method

Bidder	Cost of Items (A)	Days	Days × Multiplier (B)	Total Bid (A) + (B)
Bidder 1	$4,200,000	130	$1,300,000	$5,500,000
Bidder 2	$4,350,000	120	$1,200,000	$5,550,000
Bidder 3	$3,990,000	155	$1,550,000	$5,540,000

From Table 3.2, we find that Bidder 1 is the lowest, considering the committed number of days for construction, even though Bidder 3 is the lowest in dollar amount. The value of the contract is part "A" (cost of items) only, though—$4,200,000 in this case.

The lane renting method started in the United Kingdom, and later some government agencies in the United States started using it.[21] Like the A + B method, it adds a hypothetical monetary amount to the contractor's bid amount for the closing of road and highway lanes during construction using varying rates calculated by a formula. The main purpose is to minimize the complete or partial closure of these lanes during construction work, especially in the peak hours, and thus reduce the public's inconvenience and losses.

Liquidated damages can still apply with A + B and lane renting contracting methods if the contractor takes more days than originally committed to finish the project. In both A + B and lane renting contracting methods, with or without liquidated damages, the contractor factors all possibilities into its bid amount. Delays may and do happen, so the cost estimate has to take into consideration these possibilities and their consequences, in order to be realistic.

In addition, in some situations the contractor may have to accelerate (compress/crash) the project schedule to meet the owner's contractual requirements. More than likely, there will be a cost impact to this acceleration that must be considered when estimating or updating the project costs. The cost estimator may become a key player in providing management with information that is necessary for making decisions regarding this acceleration.[22]

21. Zohar J. Herbsman, "Lane Rental—Innovative Way to Reduce Road Construction Time," *Journal of Construction Engineering and Management* 124(5) (September/October 1998): 411–417. See also U.S. Department of Transportation: www.fhwa.dot.gov/programadmin/contracts/sep_a.cfm and Washington State: www.wsdot.wa.gov/Projects/delivery/alternative/LaneRental.htm.
22. Mubarak, *Construction Project Scheduling and Control*, 3rd ed., pp. 211–246.

Bid Shopping (Peddling)

The general contractor (GC), whether selected though competitive bidding or another process, may select its subcontractors though competitive bidding or negotiation. The GC may ask a number of subcontractors to submit price quotes before submitting its total bid to the owner. However, the winning GC may not be obligated to use the same subcontractors on the project. In this case, the GC may try to negotiate with others for a lower quote than those originally submitted with the bid. This is called *bid shopping* or *bid peddling*. In addition to being unethical, most local laws outlaw this practice.

Legal Contracts

For any valid contract, you must have:

1. *Agreement.* Offer and acceptance
2. *Consideration.* Exchange of value
3. *Capacity.* Competence of parties involved
4. *Legality.* All actions taken by all parties must be legal

Types of Contract Agreements

The written contract is often the basis of resolution of any misunderstanding, claim or dispute between the contracting parties. Writing a contract is a legal matter but can also involve technical and other specialized issues. Due to the complexity of contracts, most owners and contractors use contract templates prepared by third parties and widely used in the industry. In the United States, the most commonly used contract templates are those prepared by the American Institute of Architects (AIA) and the Associated General Contractor of America (AGC). In Europe and the Middle East, the preferred templates are those prepared by the Fédération Internationale Des Ingénieurs-Conseils or International Federation for Consulting Engineers (FIDIC) and the New Engineering Contract (NEC). The main advantage of using such forms is the confidence each party has that this is a contract prepared by an expert and neutral party and that has been used in many projects. It saves a lot of money and time as compared to hiring lawyers to write a new contract. Also, many contractors and owners are already familiar with these form contracts. This familiarity does not prevent them from making amendments and revisions to such form contracts.

The contract defines the roles and responsibilities for each of the contracting parties. Contracts differ, and so do the roles and responsibilities of the contracting parties, depending on the situation and the best interests of the contracting parties. Here are the two most common types:

1. Fixed-price contracts
   - *Lump sum.* In this type of contract, the contractor is obligated to complete and hand the project to the owner within all contract stipulations, including a total price tag that was agreed on in the contract. However, in many projects, this contract lump sum is modified due to change orders during the course of the project. Lump-sum contracts are usually used when final design documents are completed prior to signing the contract, and the contractor is familiar with type of work. In this type of contract, contractors assume the highest risk.
   - *Unit price.* This type of contract is usually used when the contractor is familiar with the type of work but—unlike in lump-sum contracts—the quantity is not exactly known. This type of agreement is common in heavy construction jobs, such as highway and railroad works, and earthwork (e.g., excavation, filling, compaction, top soil removal). Quantities have to be roughly known, though. If actual quantities vary significantly from those in the contract (usually by more than 10 percent), the contractor may be entitled to an adjustment.[23] If, however, actual quantity turned out to be much greater than originally thought, the owner may try to renegotiate the unit price with the contractor.

2. *Cost-plus-fee.* This type of contract is usually used when final design is not completed or when a contractor is not very familiar with the type of work involved in the project. Owners bear most of the risk in this type of contract. However, this risk varies with the type of cost-plus-fee contract. There are several types:
   - Cost plus a fixed (lump-sum) fee.
   - Cost plus a fixed percentage fee. This type is the least favorable to the owner because the contractor has no incentive to save money. On the contrary, the more the contactor spends, the more it makes in commission.

---

23. The main reason for this adjustment is the fact that the contractor will have to divide the fixed cost (such as mobilization, demobilization, and overhead) by the total quantity. For example, say a contractor contracted to excavate 15,000 CY of common earth at a price of $2.50/CY. Suppose the contractor's fixed cost is $5,700. The $2.50 unit price is composed of: $1.90 direct cost, $0.38 fixed cost ($5,700/15,000 CY), and $0.22 profit, all per CY. Now, if the actual quantity happened to be only 10,000 CY, the fixed cost per CY is $0.57 ($5,700/10,000 CY), and the contractor's profit shrinks to only $0.03 per CY if the unit price remains $2.50.

- Cost plus a sliding-scale fee (e.g., 18 percent for the first $500,000, 16 percent for the next $500,000, and 14 percent for all expenses above $1,000,000).

- Cost plus a fixed fee with a bonus (or penalty) for completing the project below (or above) the expected cost.

- Cost plus fee with a guaranteed maximum price (GMP). Such contracts usually contain a clause that specifies how expenses above the specified maximum are shared between the contractor and the owner.

Unbalancing Bids

Unbalancing the bid is redistributing costs among the work items involved without changing the overall total bid price. Some contractors undertake this practice in fixed-price contracts (lump sum and unit price) to increase the profit and/or shift cash flow to their advantage. It is an unethical and possibly illegal practice. It is also a risky practice that sometimes backfires on the contractor. There are three reasons why contractors unbalance bids:

1. *Unbalancing to get early money*. In this case, the contractor front-end loads its bid by deliberately and artificially inflating the cost of the early cost items and subtracting that increased amount from late cost items so the total price remains the same. This usually happens in lump-sum contracts.

2. *Unbalancing for increasing profit*. If the contractor in a unit price contract has a good reason to believe that the real quantities differ from those provided by the owner, then it can use this technique to its advantage by lowering the unit prices for items with quantities believed less than what is in the contract and increasing the unit prices for items with quantities believed more than what is in the contract. However, the contractor takes a risk if actual quantities come out different from its expectations.

3. *Unbalancing for convenience*. Sometimes contractors might want to avoid on-site arguments over the classification of soil in excavation or other type or work, so they treat all similar items (e.g., excavation) alike for bid and payment purposes.[24]

Project Delivery Methods

A *delivery method* is the approach the owner of a project uses to assign responsibility and authority among the project team members in order to

---

24. James M. Neil, *Construction Cost Estimating for Project Control* (Englewood Cliffs, NJ: Prentice Hall, 1982), pp. 305–306.

manage the entire life cycle of the project process from inception to final completion.[25] Delivery methods in construction are described as follows.

Design–Bid–Build Method

The design–bid–build method is also known as the traditional method. In this method, the owner selects the design and construction professionals in separate and consecutive processes. The main advantage to this method is that the project design has to be completed, along with its cost and time estimates, before the owner commits to project construction. The owner can later choose the contractor (construction professional, technically the *constructor* in this context) in the most suitable manner and can sign a fixed-price contract so it will get no unpleasant surprises later on. The owner can make adjustments to the design to suit the budget, time frame, or other constraints; or even to postpone or abandon the project.

There are two main disadvantages of the design–bid–build method.

1. The owner does not obtain the involvement of the construction professional (along with the associated subs and vendors) early in the design phase and thus loses potentially valuable input.

2. This method usually takes longer than other methods because of the separation of the two processes (design and construction) and the lack of the early involvement of the contractor in the project.[26]

Design–Build Method

The design–build method is called the design–construct method. In this method, the owner signs a contract with only one party to design and construct the project. This party is responsible for both the design and the construction processes of the project. In many cases, this party might be technically capable of providing only one service (design or construction) and hires or joins another entity to provide the other service, mostly through the joint-venture (JV) process.

There are two main advantages to the design–build method: First, having one point of contact and responsibility for the owner. Designer and contractor cannot play the finger-pointing game against each other because

---

25. Sometimes, especially in industrial projects, this process continues beyond final completion of the project to include operation and maintenance and perhaps upgrading and rehabilitation.
26. You can consider them three processes if you consider the bidding phase as a process, which might take a long time itself.

in this case they both represent one entity. Second, unlike the traditional method, the construction professional is involved in the project from the early stages so there will be an opportunity for better value engineering and more efficient project execution. The project execution (design and construction) is likely to be faster.

The main disadvantage of this method is the lack of clarity to the owner in terms of estimates of money and time before the owner makes a commitment. In other words, the owner signs the contract not knowing exactly the final cost or time frame. There are ways to mitigate this disadvantage by putting a cap (guaranteed maximum price, GMP) on the cost and by having the owner working together with the contractor team to bring the project to the owner's requirements including cost.

The term *turnkey project* is almost the same as design–build. In both types of projects, one party (the contractor) is responsible for the design and construction of the project. However, the contractor in the turnkey contract is also responsible for financing the project and owns the project until the project is complete and turned over to the owner. In such arrangement, the owner makes an initial down payment (deposit) and then pays the entire balance upon the contractor's completion and handover of the project, eliminating the need for progress payments.

Construction Management Method

There is a variety of forms to the construction management (CM) method, depending on the involvement and responsibility of the CM firm and other contracting parties. Basically, there are two types:

1. *Construction management at risk (CM at risk)*, where the CM acts more like a general contractor and usually hires the subcontractors to construct the project.

2. *Pure agency construction management (PCM)* (also called CM for fee or CM agency), where either the owner contracts with a general contractor to construct the project or the owner acts like a general contractor and directly hires the subcontractors.

The advantages and disadvantages of this method depend on the specific types of agreements that define the roles and responsibilities of the different contracting parties and on the phase of the project when the services of the CM firm are acquired. Obviously, the earlier the involvement of the CM firm, the better for the project.

In some projects, particularly in the design–build type, the project is accelerated in what is commonly known as the fast-track method. For more details, refer to chapter 4, "General Requirements."

Direct versus Indirect Expenses

Direct expenses are those expenses that can be linked directly to a specific work item, such as excavation for foundation; formwork for suspended slab; heating, ventilation, and air conditioning (HVAC) ductwork; or painting walls. Indirect expenses are those expenses that cannot be linked directly to a specific work item, as will be explained later.

Direct Expenses

Direct expenses include four main types:

1. *Labor.* All hourly workers' wages are considered direct expenses. Salary employees are considered indirect expenses in general—either job or general overhead. A brief labor analysis will follow (see "Indirect Expenses").

2. *Materials.* In the context of construction expenses, the materials category includes three subcategories:

   a. Installed materials such as concrete, framing lumber, bricks, or paint

   b. Installed equipment such as heat pumps, furnaces, kitchen equipment, laundry equipment, or elevators

   c. Construction materials such as formwork materials and scaffolding

Costs of installed materials and equipment (categories a and b) are charged directly in full amount. Costs of construction materials (category c) have to be depreciated or divided by the number of uses.

3. *Equipment.* In the context of construction expenses, the equipment category includes only construction equipment, such as bulldozers, cranes, and concrete pumps. A complete equipment analysis follows later in chapter 11, "Equipment Analysis."

4. *Subcontracts.* Subcontracting has increased in the past decade or two. Contractors are moving more and more toward specialization. The entire subcontract amount is considered by the general contractor (GC) as a direct expense. It is usually marked up by a certain percentage for management, additional profit, and to

compensate for some of the risk transferred from the subcontractor to the GC. To the subcontractor, this amount is composed of direct expenses (labor, materials, equipment, and—possibly—sub-subcontractors) and indirect expenses.

Indirect Expenses

Indirect expenses are those expenses that cannot be linked directly to a specific work item. There are at least five types:

1. *Job overhead*. Job overhead includes any expense that cannot be directly linked to a specific work item but can be charged to one particular project. Examples are: a temporary office (trailer), temporary utilities, bonds, permits, resident project engineer (and his or her car and cellular telephone), project secretary, clerk, and office equipment assigned to a particular project. Mobilization and demobilization can sometimes be classified as job overhead, but usually they are considered as direct expense items. Insurance policies that cover certain projects may be considered as job overhead.[27]

Contractors have to take into consideration the cost of borrowing money. This is calculated based on contractor's expected cash flow analysis, as shown in Appendix B.

2. *General overhead*. General overhead consists of any expense that cannot be directly linked to a specific work item or a particular project. This includes all expenses of the main office (structure, financing, lease, utilities, equipment, supplies, staff), taxes (income and other, excluding taxes directly linked to a specific project), advertising and public relations expenses, cost estimates for bidding, company-sponsored training programs, licenses, legal fees (except those fees tied directly to a specific project), companywide insurance, and accountant fees.

3. *Contingency*. Contingency is an amount of money set aside for events that are likely to happen but not exactly known in extent and timing.

4. *Profit*. Profit can be defined as return for taking risk, and its expected amount varies with several factors, such as risk taken and prevailing economic situation.

---

27. Usually securing permits for the permanent building are the responsibility of the design consultant or the owner, but certain permits, particularly those related to construction activities, might be the responsibility of the GC.

5. *Inflation and escalation.* Inflation represents the average change (increase or decrease) in prices of a wide collection of goods and services in the country over a period of time (usually one year). Escalation is usually used to represent the change in the price of one certain item or group of items of goods and/or services over a period of time. Both have to be considered when preparing an estimate.

Bonds and Insurance

A bond, also called *surety bond*, is an agreement whereby the bonding company (usually a specialty insurance company) guarantees to a property owner that a contractor will perform according to the contract documents. Construction surety bonds protect property owners against financial losses that could result from the default of a contractor. The most used types are bid bond, performance bond, and payment bond.

Insurance is a promise of reimbursement—by the insurance company to the insured—in the case of loss. In construction, there are many types of insurance. The most common types are workers' compensation, builder's risk, and public liability.

For more on bonds and insurance, see chapter 4, "General Requirements."

Materials Management

Labor, materials, and equipment are considered the three main resources that have to be managed in construction. Labor and equipment, as discussed in other chapters of this book, have several things in common when it comes to management. For example, their cost is subject to productivity changes, and both possibly need resource leveling. Also, unlike equipment, materials need to include a waste allowance in the calculated quantity. Materials and equipment, by contrast, are vulnerable to theft, vandalism, or damage due to multiple reasons. They need to be included in the contractor's or builder's risk insurance. They also need to have proper storage.

Materials cost can be thought of as a combination of four types:

1. *Purchase cost.* The cost of undelivered materials at the vendor's premises.
2. *Order cost.* The cost of processing, shipping, receiving, and inspecting the materials. It also includes the cost of telephone calls, faxes, administrative, and other costs pertaining to the processing of the order.

3. *Storage costs, including the additional amount of builder's risk insurance, if any, to cover these stored materials.* Storage costs also include the finance charges (or lost opportunity) accrued because of the waiting period between materials delivery and their installation (also called holding costs).[28]

4. *Unavailability (shortage) cost.* This is the potential cost (loss) to the project in case the materials were not delivered when needed.

The objective of the contractor is to minimize the total cost. It has to make decisions on when and how much of the material to order. The contractor must balance its materials purchasing philosophy between these two extreme theories:

1. *Just-in-time theory.* Materials are not delivered until the time of installation. Advantages are: low or no storage cost; low or no risk for damage, theft, or vandalism; low or no interest charges on materials held in storage. Disadvantages are: higher risk of not having the materials delivered on time, which may cause delays and increase costs; and the need for very careful coordination among vendors, labor, and management. This method may also have an advantage in the case of a change order when a certain item is changed or deleted. There is no need to return or exchange already-purchased materials.

2. *Inventory buffer.* All materials are purchased and stored at the beginning of the project. Advantages are: materials are always available for use; low order cost; possible volume discount; and possible savings if materials prices rise later on (of course, it is a loss if prices drop). Disadvantages are: high storage costs; large storage requirement; higher probability of damage, theft, or vandalism; high cost of finance charges; and larger amount of up-front money to buy the materials.

Reality, of course, is never this or that theory. It is a balance between the two theories—but this balance leans more toward one of the two theories, depending on factors such as these three:

1. *Location.* If the project is in a congested area, such as a downtown of a major city, storage space may be limited. Also, the location may be in a high-crime area that requires extra security cost. The location also influences the availability of materials. Some locations have

---

28. Contractors have to pay their bills within a certain period of time, whether they used the materials or not. Owners generally, but not always, do not pay contractors for the materials until the materials are installed and reported by the contractor and approved by the owner or his site representative.

many vendors in close proximity, so the contractor may have high confidence in being able to obtain the materials at any time.

2. *Type and quantity of materials.* If the quantity of the material to be ordered is very large, or if an item is not standard, this requires special arrangement with the vendor(s) that may influence the contractor's decision in ordering materials. Certain materials, such as ready-mix concrete, cannot be stored. Some materials require special storage arrangements (expensive materials, materials sensitive to the weather, chemicals and explosives, etc.)

3. *Criticality of schedule.* Some projects have more emphasis on finishing on time than others. Also, some activities are on the critical path while others are not. Unavailability cost depends on such factors, and the contractor plans materials delivery accordingly.[29]

Change Orders

A change order (CO) is a formal written document, signed by the owner, directing the contractor to make changes from the original contract. After its approval (by the owner), the CO becomes a part of the contract's legal documents. A change order can be used for adding, deleting, or substituting work items. The owner, design consultant (A/E), or contractor can initiate a CO, but it has to be approved by the owner. It can be initiated for reasons such as differing site conditions, design errors or omissions, or changes in the owner's requirements. In the United Kingdom, a change order is called a variation order (VO).

Technically, a change order is an alteration to the contract. If this alteration happens during the bidding period and before the bid award, it is called an addendum, and it is sent to all bidders. However, if it happens after the bid award, it is called a change order and is sent only to the contractor on the job. The change order goes to the contractor to assess the impact on the cost and schedule and then goes back to the owner for approval. The owner can approve, negotiate, or reject. It may be issued as a *change directive*, which is the same as change order but has a more decisive nature, mostly for urgency, and goes directly to execution. The contractor still assesses the impact on the cost and schedule.

One of the problems with change orders is the lack of competitiveness as the project is already under way and the work is being done by the selected contractor. This is why many change orders may escalate later

---

29. For more details on materials management, refer to CII report RS257-1, *Global Procurement and Materials Management.*

on to become disputes between owners and contractors. One indicator of success in construction projects is the absence or low rate and value of change orders. A high number of change orders mostly indicates lack of clarity of scope from the owner's side and/or repetitive change of requirements in what many professional call *scope creep*. There are many recommendations for avoiding and/or minimizing change orders, such as good planning and the involvement of the project execution teams (project management consultant, contractor) early in the process so they can help in the constructability and value engineering studies.[30] Other factors include improving the owner's organization structure and the decision-making process.

Cost Estimating, Scheduling, and Project Control

Cost estimating, as mentioned earlier, is part of project management and is integrated with all of its components. Probably the most related component to cost estimating is project scheduling: One is a prediction and measurement of money, and the other one is a prediction and measurement of time. As it is usually said, time is money. The two fields of cost estimating and project scheduling intersect in many ways. The estimator needs to know how long the project will take in order to estimate indirect costs, particularly overhead. Estimating and scheduling departments collaborate to provide information to the procurement department. The estimating department tells what resources are needed along with their quantities and prices, and the scheduling department tells when they will be needed.[31]

Both cost estimating and scheduling require breaking down the project into little elements (activities, tasks, items, etc.) that are managed and controlled easily (usually they are homogeneous, measured with one unit). However, the cost estimator and the scheduler may think differently in performing this breakdown. The cost estimator usually divides the project into items according to their nature with little or no regard to the location or chronology. So, 2×4 wood studs can be combined in one item even though they may be installed in different locations and with different timing. The scheduler's attitude toward the project breakdown is focused on the timing and location of the work item, not necessarily its

---

30. Constructability is defined as "the practice of peer reviewing the plans and specifications prior to issuing for bid with the intent of correcting errors, omissions, inconsistencies or discrepancies" (*RSMeans Illustrated Construction Dictionary*, 4th ed.). Value engineering (VE) is a science that studies the relative value of various materials and construction techniques. Value engineering considers the initial cost of construction, coupled with the estimated cost of maintenance, energy use, life expectancy, and replacement cost.

31. Read more about the relationship between cost estimating and project scheduling in Mubarak, *Construction Project Scheduling and Control* (3rd ed.), pp. 271–279.

homogeneity. For example, many schedulers combine the formwork, rebar, and concrete placement for a reinforced concrete item (e.g., suspended slab) as *FRP suspended slab*, even though formwork is measured by contact area, reinforcement is measured by weight, and concrete mix and placement is measured by volume. Concrete finishing might also be involved, which is measured by area.

Cost estimating and scheduling also tie in with other project management functions, such as procurement and accounting, and they serve as the foundation of project controls—one of the most important functions of project management.

Finally, cost estimating ties in with other project management functions, such as scope, risk, quality, safety, communications, and others. Any change in the settings or requirements of any of these functions may affect the cost of the project.

Cost Estimating Internationally

Cost estimating is the same in principle anywhere in the world; however, many factors change, such as laws and codes, rules and regulations, resource (labor, materials, and equipment) availability and prices, labor productivity, contract templates used, climate, political stability, and many other factors. Many countries do not allow foreign companies (design, construction, or project management) to operate locally unless they have a local partner. Here are some issues the cost estimator needs to be concerned with regarding international projects:

- Contract type/template, as most countries outside North America use FIDIC contracts.[32]
- Contract terms, including progress payments process and retainage.
- Availability of cost estimating database or cost references.
- Productivity of labor crews, both standard and variations with season.
- Bonds and insurance: types, cost, terms, and who provides.
- Methods of construction, including the choice of equipment and automation options.
- Unit system used: imperial versus metric system.
- Currency and exchange rate.

---

32. FIDIC is the International Federation of Consulting Engineers (the acronym stands for the French version of the name), www.fidic.org.

- Procurement process, including laws of imports and customs.
- Advance payment: In some countries, the owner has to pay the contractor 10 to 20 percent of the total contract sum in advance for the procurement of resources. That amount is recovered by deducting the same percentage from progress payments.
- Warranty on the facility imposed both by local law and by the contract: In some countries, half of the retainage money is kept with the owner until the expiration of the warranty period.
- Availability and terms of borrowing money.

It would be very risky and unwise for a cost estimator to perform estimating in a country that he or she is not familiar with, based on mere assumptions and comparisons to the home or other country. In many situations, assumptions can lead to a disaster. In one situation in which the author was involved, the estimator assumed that the cost of reinforcement steel for concrete in one country was the same as that in an adjacent country. Later, the estimator found that actual prices were about double what he assumed because that country was under international sanctions. In a different situation, the scheduler based the duration of activities on productivities in his home country. He did not realize that due to the extreme climate in the project location, productivities decline significantly, which has a direct impact on schedule and cost.

On the top of all these factors, communications in multicultural environment can be a serious challenge. Although in many situations, English is the used language (officially or unofficially), we soon find that "English is not English!" Not only do accents, pronunciation, and spelling differ, but also the meanings of many words and terms, including technical ones. It is important to make sure that what you said is what the other side understands and that everyone is on the same page.

The important lesson is that cost estimators needs to take utmost care in performing cost estimates for international projects and not make assumptions based on their background or other reference, especially when these assumptions relate to major cost items.

Chapter 3 Exercises—Set A

For solutions to the exercises in Set A, see the link to the student companion website at www.wiley.com/go/constructionestim5e.

1. Define cost, and differentiate between cost and value. Give an example.

2. Define cost estimating and quantity surveying (American term). Draw the differences between them.

3. To the quantity surveyor, doing quantity estimating to a project identical to a previous one is an easy task. This may not be the same for the cost estimator. Explain.

4. Draw similarities and show the relationship between the processes of cost estimating and scheduling. How do they meet together in "project controls"?

5. In cost estimates, rounding numbers can be a bad idea, a good idea, or a must. Explain, giving examples.

6. Why is cost estimating based on assembly takeoff generally less accurate than the one based on item takeoff?

7. Define baseline budget.

8. Moving the accuracy of the estimate from 70 to 90 percent takes a lot more effort than moving it from 0 to 70 percent. Explain.

9. What is the schedule of values? What elements does it contain? Why is it important even in lump-sum contracts? Does it have another name?

10. Private owners can award bids without competitive bidding. Government agencies, in most cases, cannot. Why? Are there exceptions? Explain.

11. An owner has complete design documents for a 20-story office building. He is asking for your opinion on the recommended type of contract and delivery method. Make recommendations and provide a justification.

12. A government agency using the A + B method on a project estimates that every day in the project is worth $20,000. There were three bidders:

	Bidder A	Bidder B	Bidder C
Bid Amount	$9,580,000	$10,560,000	$9,000,000
Duration (days)	110	95	125

Which bidder should be chosen?

13. Assume you are trying to explain the "lane renting" method to an official from the department of transportation in your state. Explain the method briefly, and list all the points the department of transportation has to consider when applying it. (You may need to do some Internet research.)

14. What type of contract do you recommend for these projects?

   A. A villa with a contractor who is experienced in this type of building.

   B. A tower building but the design in not completed yet.

   C. A road project. The total length is known approximately but not exactly.

   D. A stadium that has new, innovative features that have never been tried before.

15. The following expenses are direct expenses (D), job overhead (J), or general overhead (G)? (If expense is not clear, make an assumption and answer based on your assumption.)

   A. A pickup truck for the project manager of a building project

   B. Paint for the external walls of a building

   C. A bonus paid to all company's employees at end of the year

   D. Laptop for the cost estimator in the main office

   E. A security guard to watch out at the project site

   F. An airline ticket for the company's vice president to check the project's progress

   G. Kitchen cabinets for the new project

   H. Safety equipment for workers on a project

   I. Concrete retardant to slow curing during the summer

   J. Primavera software to schedule the company's projects

   K. Tile for the floor 7 of the building being built

   L. A bonus paid to one project's employees and workers for finishing the project ahead of schedule

   M. Scaffolding to paint the ceiling

   N. Secretary in a project office (on site)

16. What are the major factors that influence the amount (percentage) of profit a contractor desires in a project?

17. Differentiate between a constructability study and value engineering for a construction project.

18. In estimating the cost of a future project, which one is more important for the estimator: inflation or escalation? Why? Explain.

19. In balancing between just-in-time theory and inventory buffer, which theory you would lean toward if your project is:

   A. In the downtown of a major city?

   B. Away from the city and at least 100 miles from the closest city or store?

Justify your answer in each case.

20. What is bid unbalancing? Why is it risky and unethical?

21. Define *bid shopping*. What makes it unethical?

22. Write the WBS level for every item in the next table, taking in consideration that level 0 is for the entire project.

Chapter 3 Exercises—Set B

Solutions to Set B exercises are available to instructors only; see the link to the instructor's companion website at www.wiley.com/go/ constructionestim5e.

1. Put checks in the appropriate boxes in the next table.

	Statement	Approximate Estimates	Detailed Estimates
A	Requires completed design documents		
B	Used for bidding		
C	Used for feasibility studies		
D	Requires high effort		
E	Used for comparing alternative projects or designs		
F	Used for project control		
G	Quick and has high margin of error		
H	Making initial financial arrangements		
I	Its concept depends on comparing the unit cost to similar previous projects		
J	Its concept depends on actual quantities and estimated prices for the estimated project		

2. Define *cost control*.

3. Is it okay to round numbers in cost control? What is the fundamental difference between cost estimating and cost control regarding rounding numbers?

4. Create a table to compare quantity estimating (takeoff) to cost estimating. List every comparison point you can think of.

5. To the owner, the project cost is more than just the construction. Mention other possible cost items.

6. A general contractor based in New York would like to bid a project in Arizona. He has never done any work before in Arizona. Mention factors that may help him make a decision to bid or not to bid.

7. A government agency using the A+B method on a project estimates that every day in the project is worth $15,000. There were three bidders:

	Bidder A	Bidder B	Bidder C
Bid amount	$4,630,000	$5,160,000	$4,700,000
Duration (days)	120	94	110

Which bidder should be chosen?

8. If competitive bidding is one type of contract award method, what are the other two types?

9. What type of contract do you recommend for these projects?

   A. A shopping center with a contractor who is experienced in this type of buildings.

   B. A tower building for a contractor who is experienced in this type of buildings.

   C. An excavation project. The total amount of soil to be excavated is known approximately but not exactly.

   D. An opera building that has new, innovative features that were never tried before.

10. What is the rule of thumb that tells you if a cost item is a direct or an indirect expense?

11. The following expenses are direct expenses (D), job overhead (J), or general overhead (G)? (If expense is not clear, make an assumption and answer based on your assumption.)

   A. Formwork materials for a suspended slab

   B. Bulldozer to do excavation for the foundation

   C. Advertisement to congratulate the CEO for his son's wedding

   D. Printer for the project manager (on site)

   E. Overtime pay to workers to finish installing windows

   F. Party for the workers and staff to celebrate finishing the project without accident

G. Marble countertop for the kitchen

H. Safety violation for lack of temporary rail on the upper floors of the building

I. Phone bills for the main office

J. Cold water for the workers on site

K. A safety officer who serves three projects out of the company's 10 projects

12. What are the major factors that influence the amount (percentage) of contingency a contractor includes in the total cost for a project?

13. What are the advantages and disadvantages for an owner switching from the traditional delivery method to design–build method?

14. What discourages a contractor from bidding on too many projects?

15. Are the following statements true or false?

A. In the cost-plus-fee contracts, the owner assumes the biggest risk.

B. Payment bond protects the owner from financial loss resulting from failure or default of the general contractor to perform the job according to the terms and conditions of the contract.

C. To make the initial financial arrangement, an owner needs a detailed estimate.

D. Risk should be assigned to the owner when no other party can control the risk or bear the loss.

E. When a contractor is familiar with the type of project, she is more likely to sign a cost-plus-fee contract.

F. Highway/road construction is a good candidate for unit-price contracts.

G. A bid bond protects the owner from financial loss resulting from failure or default of the general contractor to perform the job according to the terms and conditions of the contract.

H. In CM-at-risk delivery methods, the CM acts almost like a general contractor.

16. Change orders can be an indicator of good or bad planning. Explain.

17. Cost-plus-fee contracts assign more risk to owners. Mention some steps owners can take in order to reduce or limit risk with this type of contract.

18. Why do some contractors unbalance their bids?

Multiple-choice questions:

19. The most important numbers in the contractor's cash flow diagram are the (check all correct answers):

    A. Total spending.

    B. Total reimbursement.

    C. Maximum debt.

    D. Cost of borrowing money.

20. Approximate estimates are used for (check all correct answers):

    A. Feasibility studies.

    B. Bidding on a project.

    C. Project control.

    D. Comparing alternative projects/designs.

    E. Making initial financial arrangements.

21. Definitive (detailed) estimates are used for (check all correct answers):

    A. Feasibility studies.

    B. Bidding on a project.

    C. Project control.

    D. Comparing alternative projects/designs.

    E. Making initial financial arrangements.

22. The methodology of approximate estimates depends on:

    A. Actual quantity takeoff from the estimated project then pricing the items.

    B. Schedule of values from the estimated project.

    C. Comparing the estimated project to similar previous projects and making adjustments.

    D. Calculating the rate of production.

23. The methodology of detailed estimates depends on:

    A. Actual quantity takeoff from the estimated project, then pricing the items.

    B. Schedule of values from the estimated project.

    C. Comparing the estimated project to similar previous projects and making adjustments.

    D. Calculating the rate of production.

24. Generally, the best source the contractor can use for estimating is:

    A. Commercial databases.

    B. Own database.

    C. Databases that come with the estimating software.

    D. Other contractors' databases.

25. One of the following is not a factor influencing the contractor's decision to bid or not to bid a project:

    A. Location of the project

    B. Size of the project

    C. Project's owner and architect/engineer

    D. Labor productivity

    E. The amount of work the contractor currently is handling

26. In normal cases, a public agency must use:

    A. Competitive public bidding

    B. Selective bidding

    C. Negotiations with bidders

    D. Direct award

27. Check the correct boxes:

No.	Bonds Statement	Bid bond	Performance bond	Payment bond
1	Required from all bidders			
2	Required from the lowest qualified bidder only			
3	Is refunded at the end of the process			
4	Guarantees that if chosen as the lowest qualifies bidder, will sign a contract with the owner			
5	Assures the owner that the contractor will pay all subcontractors, vendors, and laborers			
6	Protects the owner against liens against the project			
7	Assures the owner that the contractor will complete the project within the terms of the contract; otherwise, the surety will get another contractor to complete the project within the terms of the contract			

Chapter 4
General Requirements

CSI Division 1, General Requirements, includes all items that are not attributable to one trade or to the physical construction of the building yet are required to successfully complete the project. Items include job supervision, temporary utilities, temporary structures and services, site safety, insurance, bonds, and taxes. These items—often referred to as *project overhead*—are usually detailed in the first part of the job specifications. Additional items, though not specified, might be required to complete the project and must be estimated.

It is important to acknowledge Division 0, Procurement and Contracting Requirements. The reason this division was given the number 0 and not 1 by the Construction Specifications Institute (CSI) is because it includes tasks that precede the signing of the contract between the owner and the general contractor (GC), such as solicitation, instructions for procurement, procurement forms and supplements, contracting forms and supplements, project forms, conditions of the contract, revisions, clarifications, and modifications. Division 1, General Requirements, includes items needed during and after the stage of signing the contract between the owner and other project participants, mainly the GC. Division 0 items go to all bidding contractors, but the rest of the divisions, 1 to 49, deal mainly with the selected parties: GC and project management/construction management (PM/CM).

Estimating General Requirements

For small projects, a fixed percentage may be used for general requirements rather than estimating all items individually. Many estimators use 10 to 15 percent of the project subtotal to cover all Division 1 items. This figure can vary widely, depending on the items required to support construction operations.

When all items are to be considered individually, it is helpful to use a standard format to review a project for its general requirements. A typical form, "Project Overhead Summary," is shown below.

PROJECT OVERHEAD SUMMARY										
PROJECT							**SHEET NO.**			
							ESTIMATE NO.			
LOCATION			**ARCHITECT**				**DATE**			
QUANTITIES BY:		**PRICES BY:**		**EXTENSIONS BY:**			**CHECKED BY:**			
DESCRIPTION	QUANTITY	UNIT	MATERIAL/EQUIPMENT		LABOR		TOTAL COST			
			UNIT	TOTAL	UNIT	TOTAL	UNIT	TOTAL		
Job Organization: Superintendent										
Project Manager										
Timekeeper & Material Clerk										
Clerical										
Safety, Watchman & First Aid										
Travel Expense: Superintendent										
Project Manager										
Engineering: Layout										
Inspection/Quantities										
Drawings										
CPM Schedule										
Testing: Soil										
Materials										
Structural										
Equipment: Cranes										
Concrete Pump, Conveyor, Etc.										
Elevators, Hoists										
Freight & Hauling										
Loading, Unloading, Erecting, Etc.										
Maintenance										
Pumping										
Scaffolding										
Small Power Equipment/Tools										
Field Offices: Job Office										
Architect/Owner's Office										
Temporary Telephones										
Utilities										
Temporary Toilets										
Storage Areas & Sheds										
Temporary Utilities: Heat										
Light & Power										
Water										
PAGE TOTALS										

Project Duration

The items in Division 1 are quantified and priced *after* the estimate for Divisions 2 to 49 has been completed. This is because a full understanding of the entire scope of the project and its duration is necessary to calculate Division 1 items. Project duration can be estimated by preparing a critical path method (CPM) schedule or based on experience with past projects.

In many instances, the owner stipulates a finish date for the contractor. This is an owner's condition and does not represent a contractor's planned schedule. However, the contractor may need to accelerate its schedule in order to meet the owner's stipulated date, which is likely to increase the cost of the construction of the project.

Reference Number R012157–20 can be used to estimate construction duration if the construction completion time is not specified. Projects are listed by type and total budget in the reference. (See the reference table.)

General Requirements	R0121 Allowances

R012157-20 Construction Time Requirements

Table lists average construction time in months for different types, sizes, and values of building projects. Design time runs 25% to 40% of construction time.

Building Type	Size S.F.	Project Value	Construction Duration
Industrial/Warehouse	100,000	$8,000,000	14 months
	500,000	$32,000,000	19 months
	1,000,000	$75,000,000	21 months
Offices/Retail	50,000	$7,000,000	15 months
	250,000	$28,000,000	23 months
	500,000	$58,000,000	34 months
Institutional/Hospitals/Laboratory	200,000	$45,000,000	31 months
	500,000	$110,000,000	52 months
	750,000	$160,000,000	55 months
	1,000,000	$210,000,000	60 months

Example 1

Estimate the construction duration time for a 300,000-SF office building project with an approximate cost estimate of $35 million.

Solution

Using R012157–20 and interpolating between the 250,000- and 500,000-SF projects, the estimated time is approximately 25 months.

Example 2

Estimate the construction duration time for an industrial project with a total area of 150,000 SF.

Solution

Using R012157–20 and interpolating between the 100,000- and 500,000-SF projects, the estimated time is approximately 15 months.

Design Time and Fast-Track

These numbers are general guidelines and may vary significantly, depending on many factors. R012157–20 also estimates design time to be 25 to 40 percent of construction time.

Example 3

Estimate the design duration time for the project in Example 2.

Solution

Using R012157–20, design time runs 25 to 40 percent of construction time. This is 4 to 6 months—or about 5 months.

There are projects where construction does not wait until detailed design is 100 percent complete. In such fast-track projects, schematic design is usually done first, followed by design development. Construction overlaps with design as shown in Figure 4.1, "Fast-Track Projects." (Refer also to "When Time Matters" in Chapter 3.)

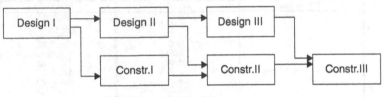

Figure 4.1 Fast-Track Projects

Project durations may also differ significantly, depending on contract obligations. Contractors might have to accelerate their schedules to meet a certain deadline, avoid liquidated damages, and/or earn an early finish bonus. This acceleration can have a significant effect on the cost.[1]

Architectural and Engineering Fees

The Reference section in RSMeans Online estimating and *Building Construction Cost Data* (*BCCD*) contains a number of charts to assist in estimating Division 1 costs. R011110–10 lists architectural fees by building type and project size. R011110–30 lists engineering fees by building type and subcontract size. Engineering fees are normally included in architectural fees.

Example 4

Estimate the architectural and engineering fees for a $2.5 million apartment complex.

Solution

Using R011110–10, architectural fees are 8.0 percent of the building cost for a $1 million project and 7.0 percent for a $5 million project.

---

1. For more details, refer to Saleh Mubarak, *Construction Project Scheduling and Control*.

Interpolating between the two percentages for a $2.5 million project, the fee is about 7.6 percent.

$$\text{Architectural fee} = \$2,500,000 \times 7.6\% = \$190,000$$

Using Reference Number R011110–30, the structural engineering fee is 1.7 percent of total project cost.

$$\text{Engineering fee} = \$2,500,000 \times 1.7\% = \$42,500$$

Performance Bond

A performance bond protects the owner from financial loss resulting from failure or default of the general contractor to perform the job according to the terms and conditions of the contract (as discussed in Chapter 3). All contractors have a *bonding capacity* that financially limits how much work they can have at the same time. For example, if the contractor has a $15 million bonding capacity and has two projects under way with a combined value of $10 million, then he or she can take on new projects with up to a $5 million value. Some surety companies limit not only the total amount of work contractors can have but also any single project as well. Sometimes when a project's value exceeds the bonding capacity of a contractor, he or she may team up with another company to raise the bonding capacity to a level allowing the team to take on the project.

Costs of performance bonds range from 0.6 to 2.5 percent of the amount of the bid. The table in note R013113–080 of the *BCCD* shows how to calculate a performance bond. Tables 4.1, 4.2, and 4.3 are similar to the one in the *BCCD* in a simpler form and using percentages. Tables 4.1, 4.2, and 4.3 show the amount of performance bond for different types of construction projects.

Table 4.1 Calculation of Performance Bond for Building Construction

Contract Amount	Performance Bond for Building Construction
up to $100,000	2.50%
$100,001–$500,000	$2,500 + 1.5% of amount over $100,000
$500,001–$2,500,000	$8,500 + 1.0% of amount over $500,000
$2,500,001–$5,000,000	$28,500 + 0.75% of amount over $2,500,000
$5,000,001–$7,500,000	$47,250 + 0.7% of amount over $5,000,000
over $7,500,000	$64,750 + 0.6% of amount over $7,500,000

Table 4.2   Calculation of Performance Bond for New Highways and Bridges Projects

Contract Amount	Performance Bond for Highways and Bridges: New
up to $100,000	1.50%
$100,001–$500,000	$1,500 + 1.0% of amount over $100,000
$500,001–$2,500,000	$5,500 + 0.7% of amount over $500,000
$2,500,001–$5,000,000	$19,500 + 0.55% of amount over $2,500,000
$5,000,001–$7,500,000	$33,250 + 0.5% of amount over $5,000,000
over $7,500,000	$45,750 + 0.45% of amount over $7,500,000

Table 4.3   Calculation of Performance Bond for Highways and Bridges Resurfacing Projects

Contract Amount	Performance Bond for Highways and Bridges: Resurfacing
up to $100,000	0.94%
$100,001–$500,000	$940 + 0.72% of amount over $100,000
$500,001–$2,500,000	$3,820 + 0.5% of amount over $500,000
$2,500,001–$7,500,000	$15,820 + 0.45% of amount over $2,500,000
over $7,500,000	$39,570 + 0.4% of amount over $7,500,000

Examples 5 to 8

Calculate the performance bond for the following jobs.

Solution

Example	Contract Type	Contract Amount	Performance Bond
5	Building	$82,000	2.5% × 82,000 = $2,050
6	Highway, new	$7,700,000	$33,250 + 0.5% × (7,700,000 – 5,000,000) = $46,750
7	Highway, resurfacing	$3,150,000	$15,820 + 0.45% × (3,150,000 – 2,500,000) = $18,745
8	Building	$25 million	$64,750 + 0.6% × (25,000,000 – 7,500,000) = $169,750

Workers' Compensation Insurance

Workers' compensation (WC) insurance covers a contractor's employees in case of personal injury or death during work or work-related illness, regardless of whose fault it was. Rates depend on three factors:

1. *Employee's trade* (carpenter, electrician, etc.). The rate for WC is directly related to the potential risk involved with that type of work. High-risk trades, such as structural steel workers (erectors, painters, welders), roofers, and wreckers, have the highest rates.

2. *Geographic location (state).* WC insurance is subject to state law and regulations. Insurance companies have the same rates within the same state but rates vary from a state to state. For example, WC insurance rates in the state of Illinois are more than six times comparable rates in the neighboring state of Indiana (see RS Means Note R013113-60).

3. *Individual employers' past safety record*, represented by its experience modification rate (EMR). The EMR ratings are calculated by the insurance industry based on the record of three years prior to the last year. For example, in 2015, the records of 2011, 2012, and 2013 are used. The average EMR for the industry is 1.0. An EMR below 1.0 indicates a good (better than average) safety record, which rewards the employer by a discount proportional to its EMR. On the contrary, an EMR greater than 1.0 represents a bad (worse than average) safety record, which increases the WC rates for that employer by a percentage proportional to its EMR.

WC insurance rates in the United States are published in the 2017 *BCCD*, page 818, divided by trade and state. They represent the contractor with an average safety record (EMR = 1.0). The U.S. average rates for different crafts are shown on the inside of the back cover. The U.S. average rate for all crafts is 12.68 percent. Note that WC insurance rates in the United States have been declining in the past few years. A possible explanation for this phenomenon is the increasing safety awareness triggered—at least in part—by the bigger role the Occupational Safety and Health Administration (OSHA) is playing in issuing heavier fines.

Example 9

What is the workers' compensation insurance rate for carpenters in Boston, Massachusetts, working on setting up formwork for concrete works? The contractor has an average safety record (i.e., EMR = 1.0).

Solution

RSMeans Reference Number R013113–60, the upper table, provides the insurance rate as a percentage of labor cost for national averages. For "Carpentry & Millwork," it ranges from 4.7 to 39.7, with an average of 13.9. Since carpentry is many types, the range in this table seems to be wide. We can take the average, a lower, or a higher number depending on the specific carpenter's job (risk involved). For our case, we will take the average, 13.9 percent.

In the lower table of the same note, we find that Massachusetts has a factor of 90, which means that WC rates there are 90 percent of national rates. For our contractor, WC rate = 13.9 × 90% = 12.51% of the basic wages of carpenters.

Example 10

Repeat the solution of Example 9 if the contractor has an EMR = 1.2.

Solution

For this contractor, WC rate = $13.9 \times 90\% \times 1.2 = 15.01\%$ of the basic wages of carpenters.

Example 11

What is the workers' compensation insurance rate for structural steel workers in Kentucky? The contractor has an EMR = 0.9.

Solution

Using RSMeans Reference Number R013113–60, Structural Steel rates range between 6.3 to 71.0, with an average of 24.4 percent of labor cost. The lower table gives Kentucky a factor of 83. For our contractor, WC rate = $24.4 \times 83\% \times 0.9 = 18.23\%$ of the basic wages of steel workers.

These examples show the amount contractors can save or pay extra for WC insurance depending on their safety record, namely EMR. Labor wages make up a significant percentage of the total construction cost, and WC insurance makes up a significant percentage of these labor wages and can reach a significant portion for a contractor with a poor safety record. (US average WC insurance rate for a skilled labor is 12.68 percent, according to the 2017 *BCCD*). RSMeans Reference Number R013113–60 shows that WC insurance makes up, on average, 6.0 percent of the total direct construction cost (before adding overhead and profit). A 20 percent savings in the WC insurance rate (with EMR = 0.80) in a $10 million project, for example, can result in a savings of nearly $100,000.[2] This is a large amount of money that not only can increase the contractor's profitability but also may allow the contractor to lower the bid and thus increase his or her chances in winning bids.

Builder's Risk Insurance

Builder's risk insurance—casualty insurance on the building during construction—is defined in R013113–40. The rates are mentioned in Means items 01 31 13.30 0020 and 01 31 13.30 0050 as 0.24 and 0.64 percent, as a minimum and maximum. If we take the average, it will be 0.44 percent. The variation in rates depends on many factors such as type of building (for example, wood frame, brick, or other),

---

2. Assuming direct cost is 80 percent of total cost, so WC insurance = $10,000,000 \times 80\% \times 6.0\% = \$480,000$. Twenty percent of that amount would be $96,000.

contract amount, deductible amount. In addition, builder's risk insurance may or may not include different types of coverage, such as fire, theft, vandalism, flood, earthquake, tornado, hurricane, and "extended coverage." Sometimes specific coverage is mandatory and regulated by the government, especially coverage for natural disasters such as floods, wind, or earthquakes. Rates differ from one location (not just the state) to another. Previous claims may have an impact on rates or even a contractor's ability to obtain a policy.

In "Installing Contractor's Overhead and Profit" (found on the inside of the back cover of the 2017 *BCCD* book), builder's risk insurance was estimated at 0.80 percent of labor cost for a national average (column C).

Example 12

Estimate the cost of builder's risk insurance for a $1,200,000 brick office building project in Boston, Massachusetts.

Solution

Taking an average of 0.44% of total cost, as in Means items 01 31 13.30 0020 and 01 31 13.30 0050:

Cost of builder's risk insurance = $1,200,000/100 × $0.44 = $5,280.

Sales Tax

The cost data in the *BCCD* and RSMeans Online estimating does not include sales taxes of any type. State or local sales taxes must be added to materials, equipment, and subcontracted work, if applicable. State sales taxes are listed in R012909–80. RSMeans does not publish local sales tax rates.

Example 13

An office building is to be constructed for a private client in New Orleans, Louisiana. The city sales tax is 5 percent. State sales taxes also apply. Calculate the total tax applicable to the project where the materials cost $2,440,000, equipment rental totals $73,000, labor costs $3,600,000, and subcontracted work totals $3,445,750.

Solution

In most situations, sales tax would not apply to labor. When estimating subcontracted work where no differentiation is made between materials and labor, assume that the work is 50 percent labor and 50 percent materials. However, the subcontractor may have already included sales tax in its bid. The estimator must also determine what types of taxes apply for equipment. For our example, we'll assume that the subcontractor did not include sales tax and that the tax rate for equipment is the same as that of materials.

Total sales tax rate	= 4% (state) + 5% (city) = 9%	
Materials sales tax	= $2,440,000 × 9%	= $ 219,600
Equipment sales tax	= $73,000 × 9%	= $ 6,570
Subcontract sales tax	= $3,445,750 × 9% × 50%	= $ 155,059
Total Tax Due		= $ 381,229

Chapter 4 Exercises—Set A

For solutions to the exercises in Set A, see the link to the student companion website at www.wiley.com/go/constructionestim5e.

1. Estimate the architectural fees for building a $15 million hospital.

2. What would the structural engineering fees for Exercise 1 be?

3. What is the performance bond for the hospital in Exercise 1?

4. An architect wants to charge an owner $340,000 for doing all architectural and engineering design for a school that is to cost around $5,000,000. Assuming the architect is average, is the owner being charged fairly?

5. What is the performance bond on the school project in Exercise 4?

6. How much do you expect the engineering fees to be for a $3.5 million parking garage?

7. What is the performance bond on a $2.75 million highway resurfacing project?

8. A water treatment plant is to be built for the city of Fort Myers, Florida. With an expected total cost of $12 million, the owner (Lee County) has decided to buy the major equipment directly. The equipment cost is $2.5 million. How much in state sales tax would the owner save by buying the equipment directly?[3]

9. What is the estimated workers' compensation insurance rate for an excavation job in North Carolina? Assume the contractor's EMR = 1.1.

10. The contractor in Exercise 9 has two crews, and each crew has four workers and a foreman who will work full time for three weeks (8 hours/day, 5 workdays/week). The hourly wage is $30/hour for the workers and $33/hour for the foreman. How much will the workers' compensation insurance cost the contractor for that job?

11. How much will the cost of workers' compensation insurance be in Exercise 10 for a contractor with EMR = 0.75? Compare the difference between the two contractors. How much would this difference be if the direct labor cost in this job is $500,000?

---

3. There are some implications, including responsibility and liability, to the owner when buying equipment. Refer to construction management books for more details

Chapter 4 Exercises—Set B

Solutions to Set B exercises are available to instructors only; see the link to the instructor's companion website at www.wiley.com/go/constructionestim5e.

1. Estimate the architectural fees for building a $3.3 million municipal building.

2. What would the structural engineering fees for Exercise 1 be?

3. What would be the mechanical and electrical engineering fees for Exercise 1?

4. What is the performance bond for the church in Exercise 1?

5. Estimate the architectural fees for a $250,000 house if the house design is:
   A. Custom built
   B. Repetitive design

6. How much do you expect the engineering fees to be for a $5 million parking garage with a complex reinforced concrete design?

7. What is the performance bond on a $4.2 million highway resurfacing project?

8. According to RSMeans data, estimate the construction time duration for a 250,000-SF office building with an estimated value of $26 million.

9. What is the estimated workers' compensation insurance rate for an earth excavation job in New Mexico (use equipment operator, medium)? Assume the contractor's EMR = 1.2.

10. The contractor in Exercise 9 has two crews (use Crew B-10S) that will work full time for four weeks (8 hours/day, 5 workdays/week). The hourly wage is $32/hour for the equipment operators and $27/hour for the laborers. How much will the workers' compensation insurance cost the contractor for this job?

11. How much will the cost of workers' compensation insurance be in Exercise 10 for a contractor with EMR = 0.80? Compare the difference between the two contractors. How much would this difference be if the direct labor cost in this job is $800,000?

Chapter 5

Adjusting RSMeans Data to Job Conditions

RSMeans's published cost data reflects average conditions: work during daylight hours, in moderate weather conditions, using average productivity. Costs are shown on a national average basis and can be adjusted, as shown in Chapter 1, by using location factors. Every project will have its own unique conditions of location, schedule, weather, and contractor. Understanding the details of how RSMeans calculates its costs will make it possible for data users to adjust costs to their specific conditions. RSMeans Online estimating has a number of functions that will be helpful in tailoring RSMeans cost data. The purpose of this chapter is to introduce techniques that can be used to customize RSMeans data to a specific set of job conditions.

Markups on Labor Cost

Project costs fall into two categories: direct and indirect costs. Direct costs are easily accounted for because they can be seen and counted (e.g., cubic yards [CY] of excavation, board feet [BF] of wood joists, or numbers of plumbing fixtures). Indirect costs, as discussed in Chapter 3 ("Direct versus Indirect Expenses," pp. 53-55), include items like the contractor's project and main office overhead, profit, unemployment taxes, and workers' compensation (WC) insurance. RSMeans uses a standard format to include the typical indirect costs as a markup on labor. These costs are detailed on the reference table "Installing Contractor's Overhead and Profit" located on the inside back cover of *Building Construction Cost Data* (*BCCD*). They can also be found in the reference section of RSMeans Online estimating by clicking on Reference Items, References, and then choosing from the drill-down menu as shown in Figure 5.1. Alternatively, if you are looking for a reference note for a specific item, click on the item so a little box opens with extra information on the item. If there is a reference note for that item, you can see the link and click it as shown in Figure 5.2. Note that labor rates are no longer part of the references. You can access them by selecting Reference Items and then Labor Rates.

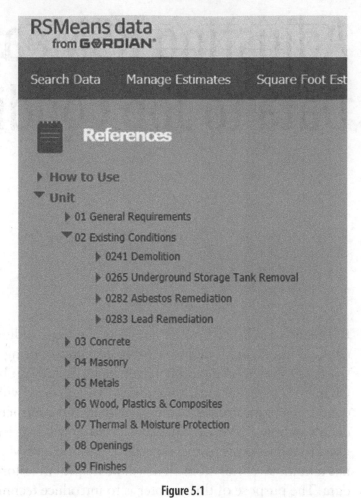

Figure 5.1

Figure 5.2

The next example demonstrates how markups on labor (often referred to as *labor burden*) are calculated.

Example 1

Find the total cost including overhead and profit for hauling 8,889 CY of soil to a remote site to be used as fill. Use the bare cost from line 31 23 16.42 1250, plus labor insurance and taxes, and 10 percent of the total direct costs for overhead and profit. The project is located in Erie, Pennsylvania (Figure 5.3).

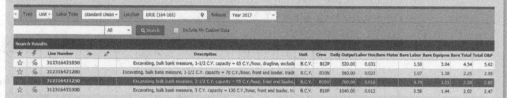

Figure 5.3

Solution

Labor burden cost (insurance and taxes) consists of four elements:

1. *Average fixed overhead.* Includes FICA (Federal Insurance Contributions Act) payment, unemployment insurance (both federal and state), and builder's risk and public liability insurance. All of these add up to approximately 18.3 percent of the base labor cost. (See the inside back cover of the *BCCD* book or Union Wages with Standard Markups on the RSMeans Online estimating, Reference Items, Labor Rates, Standard Union Labor Rates.)

2. *WC insurance.* Values range significantly depending on trade and state. (See R013113–60.)

3. *Contractor's office overhead.* This figure varies between 11 to 16 percent depending on the contractor's office expenses and total annual volume of work. The average for all trades is 13 percent of labor costs. (See Installing Contractor's Overhead, Column D, Labor Rates, Standard Union Labor Rates.)

4. *Profit.* This figure varies depending on the contractor's current situation and the job requirements. RSMeans uses a standard of 10 percent in the *BCCD*.

For this example:

WC rate, Excavation, Pennsylvania = 9.0% × 155%	= 13.95%
Average fixed overhead	= 18.30%
Overhead	= 14.00%
Profit	= 10.00%
Total labor burden cost	= 56.25%

City cost index (CCI) for Erie, Pennsylvania, Division 31, Installation (labor): 103.2% and equipment: 117.3% but the numbers in the Online Estimating (screenshot) are already adjusted for location.

Equipment bare cost	=8,889 CY×$1.51/CY	=$ 13,422
Equipment profit	=$13,422×10%	=$ 1,342
Labor bare cost	=8,889 CY×$0.79/CY	=$ 7,022
Labor burden cost	=$7,022×56.25%	=$ 3,950
Total cost, including O&P		=$ 25,737

RSMeans data provide "total cost including O&P"(overhead and profit), which is $2.85/CY.

$$\text{Total cost, including O\&P} = 8,889 \text{ CY} \times \$2.85/\text{CY} = \$25,334$$

The two approaches for calculating the total cost including overhead and profit gave close results. The difference may vary in other situations. Solving the example using the *BCCD* book numbers (national averages) with CCI numbers will probably give a different but close result. This is because CCI is implemented differently in the Online Estimating.

When using RSMeans data, it is prudent to compare them to the actual labor burden on a project. WC insurance is a major component of labor burden, and it varies widely by state, trade, and individual contractor. The rates used by RSMeans to update labor rates in the CCI tables are shown in R013113–60. If the user's actual rates vary significantly, the labor markup should be adjusted accordingly.

Interpolation between RSMeans Items

Sometimes an estimator needs an item that does not have an exact match in RSMeans Online estimating or *BCCD* book. Usually there are similar items with different parameters (such as length, width, volume, haul distance, etc.). Interpolation and extrapolation are generally acceptable. The first step is finding an item (or items) that is most similar to the designed work.

Example 2

Estimate the total direct cost for excavating 10,000 CY of common earth using 14 CY self-propelled scrapers and a pusher or dozer. The average haul distance is 2,000 linear feet (LF). Work will be in Nashville, Tennessee (Figure 5.4).

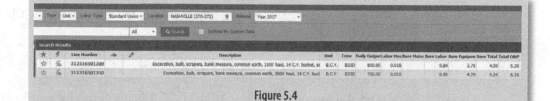

Figure 5.4

Solution

RSMeans Item No.	Description	Unit	CY/Day	Bare Costs				Total Incl. O&P
				Mat.	Labor	Equip.	Total	
31 23 16.50 1300	1500′ haul	CY	800	–	0.84	3.75	4.59	5.39
31 23 16.50 1350	3000′ haul	CY	700	–	0.95	4.29	5.24	6.16
The new interpolated item is:								
31 23 16.50 1320*	2000′ haul	CY	767	–	0.88	3.93	4.81	5.65

*This item number was created by the author. In case of interpolation, extrapolation, or just adding a new item, you may choose an appropriate item number that does not match an existing number. In the example that follows, the 1,500-foot, 3,000-foot, and 2,000-foot haul distances will be the shorter, the longer, and the new distances, respectively.

To solve for the labor, let x = the labor cost per CY:

$$\frac{\text{New distance} - \text{Shorter distance}}{\text{Longer distance} - \text{Shorter distance}} = \frac{\text{New unit cost} - \text{Shorter unit cost}}{\text{Longer unit cost} - \text{Shorter unit cost}}$$

or:

$$\frac{2{,}000' - 1{,}500'}{3{,}000' - 1{,}500'} = \frac{x - 0.84}{0.95 - 0.84}$$

Solving the equation:

$$500 / 1{,}500 = (x - 0.84) / 0.11$$
$$\text{Hence, } x = \$0.88$$

Using the same technique, we can find the equipment unit price for the new item:

$$\frac{500}{1{,}500} = \frac{y - 3.75}{4.29 - 3.75}$$
$$\text{Hence, } y = \$3.93$$

Adding up the labor and equipment costs, the total unit bare cost = \$4.81. (We also could find it using interpolation between \$4.59 and \$5.24.)

$$\text{Total bare (direct) cost} = \text{Total quantity CY} \times \text{Unit price} \times \text{CCI}$$
$$= 10{,}000 \text{ CY} \times \$4.81$$
$$= \$48{,}100$$

Using interpolation between \$5.39 and \$6.16, we find the total unit price including O&P is \$5.65.

$$\text{Total cost incl. O\&P} = 10{,}000 \text{ CY} \times \$5.65$$
$$= \$56{,}500$$

Example 3

Estimate the total direct cost for excavating 10,000 CY of common earth using 14 CY self-propelled scrapers and a pusher (dozer). Average haul distance is 6,000 LF (Figure 5.5) LF. Work will be in Nashville, TN.

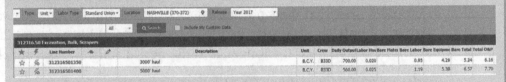

Figure 5.5

Solution

Since there is no item that exactly matches the given case, we need to extrapolate from the following two items:

RSMeans Item No.	Description	Unit	CY/ Day	Bare Costs				Total Incl. O&P
				Mat.	Labor	Equip.	Total	
31 23 16.50 1350	3000' haul	CY	700	–	0.95	4.29	5.24	6.16
31 23 16.50 1400	5000' haul	CY	560	–	1.19	5.38	6.57	7.70
The new extrapolated item is:								
31 23 16.50 1430	6000' haul	CY	490	–	1.31	5.93	7.24	8.47

Extrapolation is done as shown earlier, for labor:

$$\frac{6{,}000' - 3{,}000'}{5{,}000' - 3{,}000'} = \frac{x - 0.95}{1.19 - 0.95}$$

Thus, $x = 1.31$. Similarly, we find that the unit cost is \$5.93, \$7.24, and \$8.47 for equipment, total bare, and total including O&P.

$$\text{Total bare (direct) cost} = \text{Total quantity CY} \times \text{Unit price} \times \text{CCI}$$
$$= 10{,}000 \text{ CY} \times \$7.24$$
$$= \$70{,}240$$
$$\text{Total cost incl. O\&P} = 10{,}000 \text{ CY} \times \$8.47 / \text{CY}$$
$$= \$84{,}700$$

Note that the user has to use commonsense judgment in accepting or rejecting the answers. Also, interpolating and extrapolating may not be feasible or acceptable with materials of nonstandard sizes. The cost of a nonstandard item, if available, may be much higher per unit than the cost of the standard item.

Substituting Known Local Labor Rates

Users of the RSMeans Online estimating software can switch between union and open shop labor rates. The *BCCD* book uses union wages but there is an open-shop version of it. However, in certain situations, a contractor may have different labor rates but need to use RSMeans data for labor productivity rates, materials, and equipment costs. In other words, he or she needs to substitute the company's known labor rates for the given RSMeans labor rates.

Example 4

Estimate the bare cost for removing a $120' \times 65'$ slab on grade, 4 inches thick, reinforced with wire mesh in Reno, Nevada, based on the following local open shop wages:
Equipment operator (crane): $44/hour

Solution

Use RSMeans Line No. 02 41 16.17 0280. The crew is B13L: Two equipment operators (crane) (Figure 5.6).

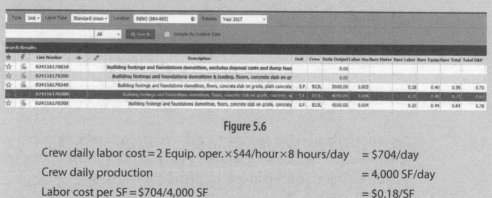

Figure 5.6

Crew daily labor cost = 2 Equip. oper. × $44/hour × 8 hours/day = $704/day

Crew daily production = 4,000 SF/day

Labor cost per SF = $704/4,000 SF = $0.18/SF

For equipment, use the number mentioned in RSMeans line: $0.49/SF, adjusted for location.

Equipment cost per SF = = $0.49/SF

Bare cost per SF = $0.18 + 0.49 = $0.67/SF

Total slab area = $120' \times 65'$ = 7,800 SF

Total bare cost = 7,800 SF × $0.67/SF = $5,226

Note: Since the labor cost is local, it is not subject to location adjustment, CCI.

Overtime Productivity Loss and Extra Pay

In the case when a worker works beyond the normal eight hours per day and five days per week, two adjustments have to be made. The first one is to make a proper reduction in labor productivity due to the fatigue factor caused by working overtime. The *BCCD* book and RSMeans Online estimating software have adjustment factors in Reference Number R012909–90. The numbers in the "Production Efficiency" columns represent the new productivity as a percentage of the normal productivity. This productivity adjustment is given for the first, second, third, and fourth week and for the average of the four weeks. The second adjustment is to be made for the pay increase during those overtime hours. Reference number R012909–90 shows the factors for two cases: pay at 1.5 and at 2 times the regular pay. The next examples explain how to make such adjustments.

Note: When dealing with materials, labor, and equipment cost; production efficiency adjustment applies to both labor and equipment, while payroll adjustment applies to labor cost only. Materials cost is not affected by overtime.

Example 5

If the contractor in Example 2 decided to work 10 hours/day, 6 days/week, what would the total bare cost and bare cost per CY be?

Solution

$$\text{Duration of the job} = \frac{\text{Total quantity}}{\text{Daily output}}$$

$$= \frac{10{,}000 \text{ CY}}{767 \text{ CY} / \text{day}} \approx 13 \text{ days}$$

Let's initially assume that with the extended hours, the contractor can finish the job in 12 days, which is two weeks (based on six workdays/week). Using the overtime table, R012909–90:

$$\text{Production efficiency (average two weeks)} = \frac{(95\% + 90\%)}{2} = 92.5\%$$

Payroll cost factor $= 116.7\% = 1.167$ (from the same table)

Normal bare cost per CY (from Example 2) is $0.88 for labor and $3.93 for equipment.

The easy way of solving this example is:

$$\text{Total bare unit cost} = M + \frac{L \times \text{Payroll factor} + E}{\text{Production efficiency}}$$

where M, L, E are unit cost for materials, labor, and equipment, respectively.

$$\text{Total bare unit cost} = 0 + \frac{\$0.88 \times 1.167 + \$3.93}{92.5\%} = \$5.36 / CY$$

$$\begin{aligned}
\text{Total bare cost} &= \text{Total quantity} \times \text{Bare cost per CY} \times \text{Location factor (CCI)} \\
&= 10,000 \, CY \times (\$0.88 \times 1.167 + \$3.93) / 0.925 \\
&= \$53,589 \approx \$53,600
\end{aligned}$$

For a more detailed solution:

$$\text{Normal hourly production} = \frac{767 \, CY/day}{8 \, hours/day} = 96 \, CY/hour$$

$$\begin{aligned}
\text{New daily production} &= 96 \, CY/hour \times 10 \, hours/day \times 92.5\% \\
&= 888 \, CY/day
\end{aligned}$$

If we go to the crew used, B-33D, and separate bare labor from bare equipment, we get:

$$\begin{aligned}
\text{Crew daily labor cost} &= \$428.40 + \$156.60 + \$107.10 = \$692.10 \\
&\quad \text{(National average)}
\end{aligned}$$

$$\begin{aligned}
\text{Crew daily equipment cost} &= \$2,429.00 + \$475.00 &= \$2,904.00 \\
&\quad \text{(National average)}
\end{aligned}$$

To make the proper adjustment, we have to multiply both labor and equipment by (new hours/day over normal hours/day) as well as multiply labor cost by payroll factor. Also, make CCI adjustments for Nashville, TN.

$$\text{New crew daily labor cost} = \$692.10 \times 10 / 8 \times 1.167 \times 95.9\% = \$968.21$$

$$\text{New crew daily equipment cost}^{[1]} = \$2,904 \times 10 / 8 \times 103.4\% = \$3,753.42$$

$$\text{New crew daily total bare cost} = \$968.21 + \$3,753.42 = \$4,721.63$$

$$\begin{aligned}
\text{New bare cost per CY} &= \frac{\text{New crew daily total bare cost}}{\text{New crew daily production}} = \frac{\$4,722/day}{888 \, CY/day} \\
&= \$5.32 / CY
\end{aligned}$$

---

1. In the equipment calculation, we made a simplifying assumption that equipment cost will be the same per hour. In reality, equipment cost per hour in case of overtime may slightly decrease, because the rent cost per hour drops slightly. (Refer to RSMeans Reference Table R015433-10.) The cost for operating (fuel and oil) per hour stays the same.

This is close to the result we obtained earlier, $5.36.

For calculating the total cost including O&P, we have to use the detailed approach:

Crew daily labor cost	$= \$648 + \$240 + \$162$ (Before adjustment)	$= \$1050$
Crew daily equipment cost	$= \$2,671.90 + \$522.50$ (Before adjustment)	$= \$3,194.40$
New crew daily labor cost	$= \$1050 \times 10/8 \times 1.167 \times 95.9\%$	$= \$1,468.89$
New crew daily equipment cost	$= \$3,194.40 \times 10/8 \times 103.4\%$	$= \$4,128.76$
New crew daily cost incl. O&P	$= \$1,468.89 + \$4,128.76$	$= \$5,598$

$$\text{New cost per CY incl. O\&P} = \frac{\text{New crew daily cost incl. O\&P}}{\text{New crew daily production}}$$

$$= \frac{\$5,598/\text{day}}{888 \text{ CY/day}} = \$6.30/\text{CY}$$

Total cost incl. O&P	$= 10,000 \text{ CY} \times \$6.30$	$= \$63,000$

Note: In applying the adjustment factors (productivity; 92.5%, and payroll; 1.167) to the total cost including overhead and profit, there was a small error that increased the labor cost (overestimate). This is because we applied these factor to labor cost including all labor add-ons. Most of these add-ons do not increase at the same rate as the direct payroll does or they are capped. However, it is believed that this error is relatively small and safe to ignore.

Example 6

The contractor in Example 2 believes that, due to adverse conditions, the productivity will drop by 20 percent. (No overtime is needed.) What is the total bare cost?

Solution

Since the cost of these items consists of labor and equipment only (no materials), we can simply adjust the cost in inverse proportion to the productivity change.

Bare cost per CY	$= \dfrac{\$4.81/\text{CY}}{(1 - 0.20)}$	$= \$6.01$
Cost per CY incl. O&P	$= \dfrac{\$5.65/\text{CY}}{(1 - 0.20)}$	$= \$7.06$
Total bare cost	$= 10,000 \text{ CY} \times \$6.01/\text{CY}$	$= \$60,100$
Total cost incl. O&P	$= 10,000 \text{ CY} \times \$7.06/\text{CY}$	$= \$70,600$

Note: Remember that productivity changes do not affect materials cost.

Effect of Inflation/Cost Escalation

Contractors are often required to estimate the cost of future projects, which may be planned to start immediately or later in the future. Construction costs typically change over time. Some items' cost increases may be close to national inflation numbers, while others may be significantly higher. Other costs may even decrease. As a realistic expectation, contractors usually add an inflation/escalation allowance as a percentage of the total cost. This allowance may be based on trends in the past few years, expectations of future economy based on current events, or sheer instinct. No matter what the method is, it cannot be completely accurate, but it should be reasonable. This topic was discussed in more detail in Chapter 3.

The inflation/escalation allowance is factored into the estimate using the compound interest equation:

$$F = P \times (1+i)^n$$

where
F = Future estimated cost
P = Present estimated cost
i = Expected annual inflation/escalation rate
n = Number of years between the estimate and expected midpoint of the considered project

Good references for price changes include *RSMeans Construction Cost Indexes* (quarterly publications) and *Engineering New Record's* (*ENR*) quarterly cost reports. The *BCCD* and RSMeans Online estimating software include historical cost indexes that allow the estimator to adjust for cost changes over the years. Inflation over a certain period of time can be calculated using the equation just provided, with F, P, and n as known values and i as the unknown. (More on time value of the money is covered in Appendix B.)

Here is another important tip in estimating n, the time period in the equation. Since the database the estimator is using may not be current, then n must start at the date of the prices (the database) and not the time of the estimate. The inflation for the period n will be include a deterministic part (during n_1) and an estimated part (during n_2) (Figure 5.7).

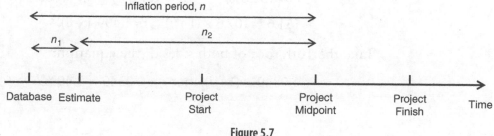

Figure 5.7

Example 7

What was the average cost escalation for the following periods?

- a. 2010–2015
- b. 2005–2010
- c. 1995–2005
- d. 1995–2000
- e. 1970–1980

Solution

Use the historical cost indexes with January 1, 1993, as a base. To find the historical cost indexes, go to: Reference Items → City Cost Indexes → Historical Cost Index.

a. For the five-year period 2010 to 2015:

$$F = 206.2\%, \qquad P = 183.5\%, \qquad n = 5$$
$$206.2 = 183.5 \times (1 + i)^5, \text{ or } 1.124 = (1 + i)^5$$

Take the 5th root of both sides of the equation:

$$1.0236 = (1 + i) \text{ then } i = 0.0236 = 2.36\%$$

Historical Indexes may slightly differ in the BCCD between the print book and the online materials because one is based on prices in January and the other in July of the same year.

b. For the five-year period 2005 to 2010:

$$F = 183.5\%, \qquad P = 151.6\%, \qquad n = 5$$
$$183.5 = 151.6 \times (1 + i)^5, \text{ or } 1.210 = (1 + i)^5$$

Take the 5th root of both sides of the equation:

$$1.0389 = (1 + i) \text{ then } i = 0.0389 = 3.89\%$$

Note that inflation was negative from 2008 and 2009, but very small in amount.

c. For the 10-year period 1995 to 2005:

$$F = 151.6\%, \qquad P = 107.6\%, \qquad n = 10$$
$$151.6 = 107.6 \times (1 + i)^{10}, \text{ or } 1.409(1 + i)^{10}$$

Take the 10th root of both sides of the equation:

$$1.0349 = (1 + i), \text{ then } i = 0.0349 = 3.49\%$$

d. For the five-year period 1995 to 2000:

$$F = 120.9\%, \qquad P = 107.6\%, \qquad n = 5$$

$$120.9 = 107.6 \times (1 + i)^5, \text{or } 1.1236 = (1 + i)^5$$

Take the 5th root of both sides of the equation:

$$1.0236 = (1 + i), \text{then } i = 0.0236 = 2.36\%$$

e. For the 10-year period 1970 to 1980:

$$F = 62.9\%, \qquad P = 28.7\%, \qquad n = 10$$

$$62.9 = 28.7 \times (1 + i)^{10}, \text{or } 2.192 = (1 + i)^{10}$$

Take the 10th root of both sides of the equation:

$$1.0816 = (1 + i)$$

$$i = 0.0816, \text{or } 8.16\%$$

As the previous example shows, we find that inflation fluctuates, and there is no scientifically accurate way of predicting it.

Example 8

A contractor needs to excavate 12,000 CY of heavy soil using a track-mounted 3 CY front-end loader. The job will take place in Los Angeles, California, in early 2019. What is the expected bare cost and the total cost including O&P?

Solution

Using RSMeans Lines No. 31 23 16.42 1300 and 31 23 16.42 4100, the combined items are already adjusted for Los Angeles and can be summarized as follows:

RSMeans Item No.	Description	Crew	Unit	CY/Day	Bare Costs				Total Incl. O&P
					Mat	Labor	Equip	Total	
31 23 16.42 1300	Excavation, bulk bank measure, 3 CY capacity = 130 CY/hour, front-end loader, track mounted, excluding truck loading	B10P	CY	1040	–	0.61	1.25	1.86	2.30
31 23 16.42 4100	Excavating, bulk bank measure, for heavy soil or stiff clay, add							60%	60%
31 23 16.42 1300/4100	Item 31 23 16.42 1300 modified for heavy soil	B10P	CY	650	–	0.98	2.00	2.98	3.68

The daily production for heavy soil was roughly calculated as shown next:

$$\text{Cost per CY} = \frac{\text{Crew daily cost}}{\text{Crew daily production}}$$

Since the crew daily cost is constant, the cost per CY will be inversely proportional to the crew daily production:

$$\frac{\text{Cost per CY1}}{\text{Cost per CY2}} = \frac{\text{Crew daily production 2}}{\text{Crew daily production 1}}$$

$$\text{Crew daily production (heavy soil)} = \frac{1.67}{2.67} \times 1040 = 650 \text{ CY}$$

These equations are good for both bare cost and cost including O&P. For inflation, assume a 5 percent annual inflation rate for the next two years.

$$\begin{aligned}
\text{Present total bare cost} &= \text{Total quantity} \times \text{Unit price} \\
&= 12,000 \text{ CY} \times \$2.98/\text{CY} \\
&= \$35,760 \\
\text{Future total bare cost} &= \$35,760 \times (1+5\%)^2 \\
&= \$39,425 \approx \$39,400 \\
\text{Present total cost incl. O\&P} &= 12,000 \text{ CY} \times \$3.68/\text{CY} \\
&= \$44,160 \\
\text{Future total cost incl. O\&P} &= \$44,160 \times (1+5\%)^2 \\
&= \$48,686 \approx \$48,700
\end{aligned}$$

Example 9

A large construction project was estimated to cost $43,500,000 in January 2017 dollars. How much do you expect it to cost if it started in February 2019 and will be completed in November 2021?

Solution

First, let's assume a project's midpoint of July 2020. This means $n = 3.5$ years (the time distance between the date of the cost database and the project's midpoint).

Just to illustrate the impact of a slight change in the assumption of inflation, we will solve this problem twice: once for a liberal estimated inflation of 3 percent and another for a conservative estimated inflation of 5 percent.

$$\text{For } i = 3\%: F = \$43,500,000 \times (1+3\%)^{3.5} = \$48,241,000$$

$$\text{For } i = 5\%: F = \$43,500,000 \times (1+5\%)^{3.5} = \$51,600,000$$

A balanced cost would be $50 million.

Unit Consistency

Units have to be matched carefully. In some cases, two cost items in the same job may be estimated with different units. For example, in the case of site demolition, demolition cost is given in $/CF (cubic feet), while hauling and dumping debris, which is found in the same RSMeans classification, is given in $/CY (cubic yards).

Example 10

Estimate the total bare cost for demolition of an 8″-thick brick wall, 8′ high and 120′ long, in Durham, North Carolina. Debris will be disposed of at a site 4 miles away.

Solution

For demolition, use these RSMeans items:

RSMeans Item No.	Description	Unit	Bare Costs				Total Incl. O&P
			Mat.	Labor	Equip.	Total	
02 41 13.30 1200	Demolition, wall, brick, solid	C.F.	–	2.49	1.80	4.29	5.78
02 41 13.30 4500	Disposal, up to 5 miles	C.Y.	–	4.40	10.05	14.45	17.69

$$\text{Area of the wall} = \text{Length} \times \text{Height} = 120' \times 8' = 960 \text{ SF}$$
$$\text{Volume of the wall} = \text{Area} \times \text{Thickness} = 960 \text{ SF} \times 8 \text{ in}/12 \text{ in}/\text{ft}$$
$$= 640 \text{ CF}$$
$$\text{Volume in cubic yards} = 640 \text{ CF}/27 \text{ CF}/\text{CY} = 24 \text{ CY}$$

$$\text{Bare cost for demolition} = \text{Volume} \times \text{Unit cost in CF}$$
$$= 640 \text{ CF} \times \$4.29 = \$2,746$$
$$\text{Bare cost for hauling} = \text{Volume} \times \text{Unit cost in CY}$$
$$= 24 \text{ CY} \times \$14.45 = \$347$$
$$\text{Total bare cost} = \$2,746 + \$347 = \$3,093 \approx \$3,100$$

Note: Volume of the wall debris to be hauled will be greater than 24 CY due to the voids in the debris, but we will ignore this issue here for simplicity.

Adding Custom Data

RSMeans Online estimating allows the user to add custom lines. In Example 2, we interpolated between two lines to establish a cost for hauling a 2,000-foot distance. We can either add this item to the spreadsheet outside of RSMeans Online estimating or add it to the estimate using the "Insert custom line" function. But in order to add a custom line item, you need to open an estimate. Here's an example:

Example 11

Enter the line of data created earlier in Example 2—that is, 31 23 16.50 1320—into the RSMeans Online estimate as a custom line.

1. Open the estimate, click on the "Insert custom line" button. The "Insert Line Item" will pop up (Figure 5.8).

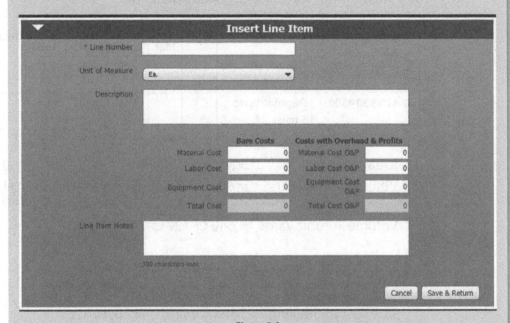

Figure 5.8

2. Enter the data for the custom line 31 23 16.50 1320:

   • Description: Excavation, bulk, scrapers, bank measure, common earth, 2000′ haul, 14 CY bucket, self-propelled scrapers, 1/4 push dozer

   • Unit of measure (UOM): CY

- Quantity: 1
- Material cost: $0.00
- Labor cost: $0.88
- Equipment cost: $3.93
- Material cost O&P: $0.00
- Labor cost O&P: $1.33
- Equipment cost O&P: $4.32

3. You can add a note: "Interpolation between lines 31 23 16.50 1300 and 1350 for a 2000' haul."

4. When all data are entered as in Figure 5.9, click Save & Return.

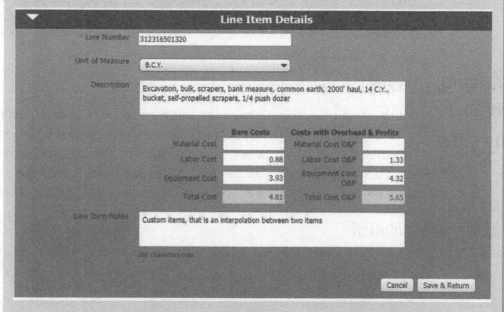

Figure 5.9

5. The "Insert custom line" windows closes authomatically and you return to the estimate. The new item is there with 0 quantity. Add the quantity 8889 (unit is CY) and hit the enter key on the keyboard. You will see the cost numbers change to reflect the quantity. (See Figure 5.10.)

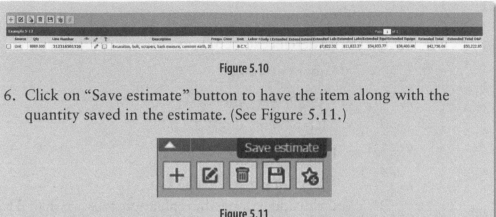

Figure 5.10

6. Click on "Save estimate" button to have the item along with the quantity saved in the estimate. (See Figure 5.11.)

Figure 5.11

Note: The line item has been added to the estimate. The source is listed as Custom (the pencil icon). It will not be resident in the database so the item can only be used in this estimate. You can edit it by double-clicking the item number as explained in the next example.

Adding, Changing, or Deleting Costs

Example 12

In the previous example, no material costs were included. Assume that the cost of diesel fuel has skyrocketed recently far beyond the operating cost allowed in the hourly cost used by RSMeans. The contractor has figured that a surcharge of $0.50 per CY will be needed to cover the increased fuel cost. Revise line 31 23 16.50 1320 to include a material cost for fuel.

Solution

Double-click on the line number 31 23 16.50 1320 (custom line) in the estimate to bring up the Line Item Details screen, as in Figure 5.12.

On the Material Cost line, add $0.50.

On the Material Cost O&P line, add $0.55.

Click Save & Return.

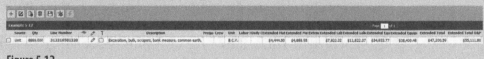

Figure 5.12

You can add, delete, or revise material, labor, or equipment costs in this manner, bare or with O&P.

Create an Estimate Summary—Reports

RSMeans Online estimating has several reporting options that give summary or detailed estimate, or certain customization options to the user. Figures 5.13A and B show the options of the Report function. Figure 5.13C shows the options of the Advanced Reporting function.

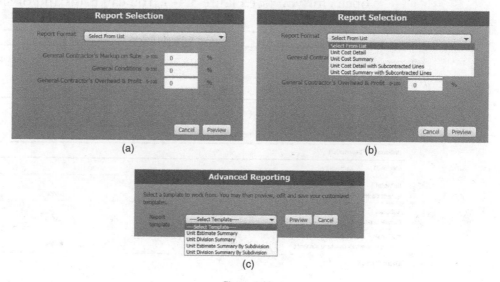

Figure 5.13

Example 13

Create an estimate summary for the estimate in Example 12. Add the following amounts:

General Contractor's Markup on Subs

Solution

1. On the Estimate toolbar, click on Estimate Action and choose Reports (as shown earlier in Figure 5.13A)

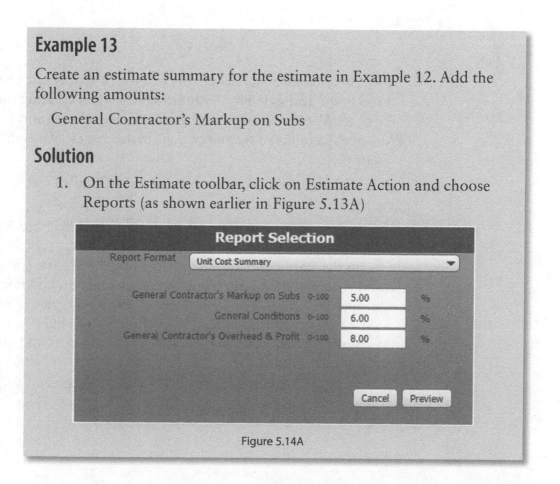

Figure 5.14A

2. In the Report Selection dialog box, choose Unit Cost Summary and enter the specified percentages. Click Preview and the report opens for you (see Figure 5.14). You can print and/or save it.

Cost Estimate Report

Date: 11/06/2017

1234 Main Highway
Nashville, TN

Example 5-12
Year 2017
Unit Summary Report
Prepared By: Saleh Mubarak   SALAH MUBARAK

Division	Description		Total
Division 31	Earthwork		$55,111.80
Subtotal			**$55,111.80**
General Contractor's Markup on Subs		5.00%	$0.00
Subtotal			**$55,111.80**
General Conditions		6.00%	$3,306.71
Subtotal			**$58,418.51**
General Contractor's Overhead and Profit		8.00%	$4,673.48
Grand Total			**$63,091.99**

RSMeans data
from GORDIAN

1

Figure 5.14B

3. You cannot change the header in this type of report but you can do so in the Estimate Header Information when creating the estimate. Also, you can control your general information using Manage Accounts in the upper right-hand corner (Welcome <your name>).

4. Repeat the step above, choosing Unit Cost Detailed (see Figure 5.15).

5. If you like to edit your report, you can use the Advanced Report option that allows you to build almost any report you can think of with the RSMeans data. In order to edit the report, you need to Preview and then Edit. Note that in selecting and adding Markup items, you cannot edit existing ones but you can create similar items with the amounts / percentages you like. You can download the report in Word, Excel, or PDF format and save it. (see Figure 5.16).

Another option is to export the estimate to Excel and edit/maniplulate it as you like.

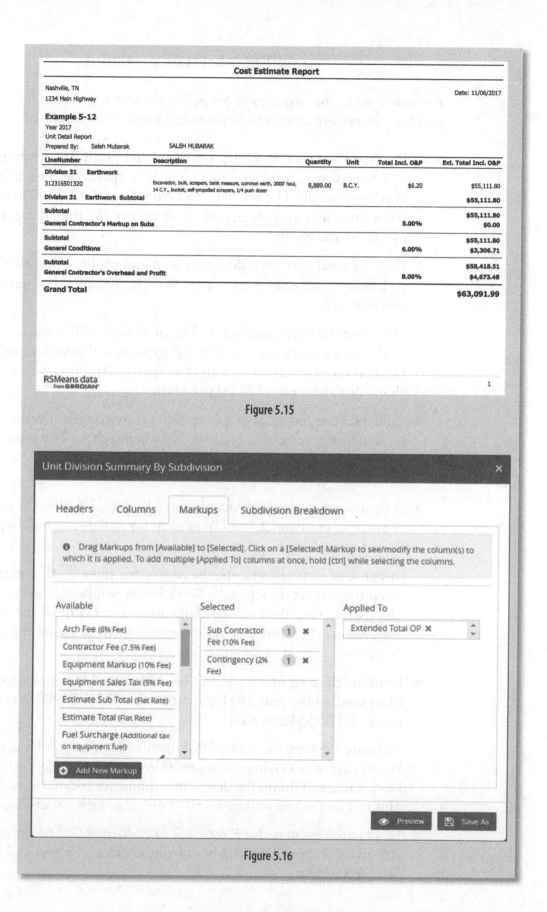

Cost Estimate Report

Nashville, TN
1234 Main Highway

Date: 11/06/2017

Example 5-12
Year 2017
Unit Detail Report
Prepared By: Saleh Mubarak   SALEH MUBARAK

LineNumber	Description	Quantity	Unit	Total Incl. O&P	Ext. Total Incl. O&P
Division 31 Earthwork					
312316501320	Excavation, bulk, scrapers, bank measure, common earth, 2000' haul, 14 C.Y., bucket, self-propelled scrapers, 1/4 push dozer	8,889.00	B.C.Y.	$6.20	$55,111.80
Division 31 Earthwork Subtotal					**$55,111.80**
Subtotal					$55,111.80
General Contractor's Markup on Subs				5.00%	$0.00
Subtotal					$55,111.80
General Conditions				6.00%	$3,306.71
Subtotal					$58,418.51
General Contractor's Overhead and Profit				8.00%	$4,673.48
Grand Total					**$63,091.99**

RSMeans data
from GORDIAN

1

Figure 5.15

Unit Division Summary By Subdivision   ×

Headers   Columns   **Markups**   Subdivision Breakdown

ⓘ Drag Markups from [Available] to [Selected]. Click on a [Selected] Markup to see/modify the column(s) to which it is applied. To add multiple [Applied To] columns at once, hold [ctrl] while selecting the columns.

Available

Arch Fee (6% Fee)

Contractor Fee (7.5% Fee)

Equipment Markup (10% Fee)

Equipment Sales Tax (5% Fee)

Estimate Sub Total (Flat Rate)

Estimate Total (Flat Rate)

Fuel Surcharge (Additional tax on equipment fuel)

⊕ Add New Markup

Selected

Sub Contractor   ① ✖
Fee (10% Fee)

Contingency (2%   ① ✖
Fee)

Applied To

Extended Total OP ✖

👁 Preview   💾 Save As

Figure 5.16

Chapter 5 Exercises—Set A

For solutions to the exercises in Set A, see the link to the student companion website at www.wiley.com/go/constructionestim5e.

1. Find the total cost, including overhead and profit, for erecting 175 tons of structural steel in a nine-story apartment building in Washington, DC. Use the bare cost from line 05 12 23.77 0400 plus labor insurance and taxes, and 10 percent of the total direct costs for overhead and profit.

2. Find the total cost, including overhead and profit, to install a pedestrian bridge, steel arch span, 8′ wide, 90′ span. Use national average costs.

3. Calculate the bare cost for erecting an 8′ high, 120′ long, 2 × 4 wood stud interior partition wall with ½″ gypsum wallboard, taped, and finished on two sides. The wall is in Tucson, Arizona, where open shop carpenters earn $32.00 per hour.

4. Calculate the cost of laying 6-oz. polypropylene stabilization fabric on a highway project in Nashua, New Hampshire. The project will require 10,000 SY of fabric. Laborers will earn $31.00/hour, and equipment operators will earn $40.00/hour.

5. A contractor is using RSMeans item 31 23 16.30 2000 to calculate the cost to excavate 2,500 CY of rock for a high-rise building in downtown Memphis, Tennessee.

   Because of time constraints, the contractor must work overtime to keep the project on schedule. Work hours will be 9 hours/day, 6 days/week. Overtime will be compensated at 1.5 times the standard pay. Adjust the labor cost and productivity to calculate the total bare cost.

6. Estimate the cost of a project to be constructed in Annapolis, Maryland, in the year 2021. An estimate of $5,575,000 was done using 2017 RSMeans data.

7. Estimate the total including O&P for the demolition of a gypsum board suspended ceiling on suspension system. Include the cost to load, haul, and dump the debris in a dumpster located approximately 100 feet away. The ceiling is 120′ × 60′. Use national average costs.

8. Display the estimate for Exercise 7 in RSMeans Online estimating Advanced Reporting, using both Unit Summary Report and Unit Detailed Report.

9. Display the estimate for Exercise 7 in RSMeans Online estimating Advanced Reporting, using Unit Estimate Summary. Add:

> Contingency (2% Fee on Extended Total)
>
> GC O&P (15% Fee on Extended Total)
>
> Labor Markup (57% Fee on Labor)
>
> Equipment Fuel Surcharge (5% of equipment cost, this is a custom item)

10. Repeat Exercise 9 by exporting the estimate to Excel and then adding the four items mentioned.

11. Do an estimate for Example 1 using RSMeans Online estimating report. (Note: The value may not exactly match the value calculated in Example 1 because RSMeans Online estimating uses a more detailed location factor table than the one shown in the *BCCD*.)

Chapter 5 Exercises—Set B

Solutions to Set B exercises are available to instructors only; see the link to the instructor's companion website at www.wiley.com/go/constructionestim5e.

1. Find the total cost, including overhead and profit, for erecting 85 tons of structural steel in a one-story industrial building, steel bearing with bolted connections, in Roanoke, Virginia.

2. Find the total cost including overhead and profit to install steel chain-link fence, vinyl coated, 2″ mesh, 9 gauge. This fence will be 560 LF long and 4′ high, and will be located in Norman, Oklahoma.

3. Calculate the bare cost for erecting interior partition walls, 8′ high, 322′ long. The blocks will be 6″ thick, lightweight, hollow, 2000 psi, not reinforced. The walls will be located in New Orleans, Louisiana, where an open shop mason earns $38.00 per hour and the helper earns $32.00 per hour.

4. If the contractor in Exercise 3 needs to finish this job in five days with the same crew (as specified in RSMeans), how many hours/day should the crew be doing? Use RSMeans Note R012909–90.

5. Calculate the cost of laying a ¾″ stone base course for a roadway in Fairfax, Virginia. This base course will be compacted, 6″ deep, 25′ wide, and 1 mi long.

6. A project was estimated at $7,550,000 based on 2017 RSMeans data. How much would you estimate its cost if it is to be constructed in the year 2021? Assume an annual inflation rate equal to the one in between 2005 and 2015.

7. Estimate the total including O&P for the demolition of a building with ceilings and partitions containing asbestos. The ceilings, 2,700 SF, include suspension system and plaster and lath. The partitions, 9,600 SF, have plaster and lath and studs. Use national average costs.

8. Display the estimate for Exercise 7 in RSMeans Online Estimating report. Use your initials to name your construction company.

9. Display the estimate for Exercise 7 in RSMeans Online Estimating Advanced Report, Unit Estimate Summary.

Add the following:

Contingency (2% Fee on Extended Total)

GC O&P (15% Fee on Extended Total)

Sub Contractor Fee (10% Fee on Extended Labor)

Fuel Surcharge (5% of Extended Equipment) this is a custom item that you need to add 796.8745.

10. Repeat Exercise 9 using the Export to Excel function and then adding the mentioned items in Excel.

Chapter 6

Concrete
Division 3

Types of Concrete

Concrete is the most extensively used material for construction in the world. It has many advantages: strength, flexibility in forming, durability, and high resistance to fire. In many situations, it is the most economical choice.

Concrete can be precast or cast in place, though cast-in-place concrete is most commonly used. Ready-mix concrete is prepared by professional companies in a controlled environment, resulting in a high-quality mix. Site (job) mix is used in cases where the large quantity of concrete needed justifies the placement of a mix plant on the site or when the project site is far away from ready-mix plants.

Precast

Precast concrete is made under controlled conditions in special plants using mechanized automated systems and a high-standard quality assurance/quality control (QA/QC) program, resulting in high-quality concrete, both in shape and in strength. Since precast concrete requires special handling and transportation, it is usually delivered—and sometimes installed—by the manufacturer.

Cast in Place

Formwork: Building and erection of formwork is usually the most consuming step, both in time and cost, in the reinforced concrete process. Technically, *formwork* is the material that holds the freshly placed concrete. It is measured by the contact area (square feet of contact area, SFCA). *Falsework* is a term that includes formwork plus other supporting elements, such as shores, scaffolding, and lateral bracing.

The cost of formwork varies significantly from one job to another, depending mainly on the complexity of design. Many types of materials are used for formwork. Lumber is most often used in the United States. Other materials include plywood, steel, aluminum, PVC, fiber, fiberglass, and rubber. Accessories used to keep the forms in place include ties, clamps, shores, scaffolding, screw jacks, and air pumps. The cost of formwork varies with the number of uses, among other factors. Contractors usually select the type that gives them the least cost per SFCA. Thus, for fewer than five uses, lumber is the most economical. For a large number of uses, steel is a good option. Steel and other expensive forms may be rented or leased.

Reinforcement: *Reinforcement* (reinforcing, rebar, or steel bars) refers to the operation of adding steel to the concrete to help resist primarily tensile and shear stresses and, occasionally, compressive stresses. Reinforcement is available in these shapes:

- *Regular reinforcing bars* (usually called *rebar*) are available in sizes ranging from number 3 to number 18. The designation number (size) for numbers 3 through 8 represents the diameter in increments of ⅛ of an inch. For example, size number 5 has a diameter of ⅝ inch. All bars' designation numbers, diameters, weight per lineal foot, and other information are displayed in Reference Table R032110–10 in the *RSMeans Building Construction Cost Data* (*BCCD*) and RSMeans Online estimating, Reference Items, References.

- *Welded wire fabric* (wire mesh) is supplied in rolls or sheets. The designation number is composed of two pairs of numbers. The first represents the spacing of the wires (in both directions), and the second represents the gauge number (in both directions). As the gauge number increases, the actual wire size (diameter) decreases, and vice versa. Welded wire fabrics usually have smaller size bars (wires) than reinforcing bars. They are most often used for slabs on grade and occasionally for small, suspended slabs.

- *Regular stressing tendons* (cables) are composed of multiple strands of very high strength steel (about 273 KSI [kilopound per square inch]) and used for prestressed concrete.

- *Reinforcing steel* is most commonly grade 60. Grades 40 and 75 are also available. These numbers represent the yield stress in kips per square inch. Steel grades for stressing tendons, used for prestressed concrete, are much higher.

Placement Methods: To estimate concrete installations, one must include the mix and its placement. The concrete mix should comply with the minimum compressive strength required. Placement can be done in one of three methods:

1. *Direct chute*. This is the least costly method, used only for concrete members at or below grade level. The mix is dropped (poured) directly from the mix truck through a chute to the desired placement location by gravity force, often with help from laborers pushing the concrete down the chute.

2. *Crane and bucket*. This method is used for higher locations or locations that cannot be reached with the mix truck. It may also be used when the contractor has a crane already on site. In small jobs, buckets may be carried by laborers and handled manually.

3. *Pumping*. This method is used for higher locations or locations that cannot be reached with the mix truck. Pumping concrete uses different pumps and attachments depending on factors such as quantity of concrete, location and elevation of concrete member, and thickness of the concrete mix.

   In many jobs, both crane and bucket and pumping are viable methods, and the contractor chooses one method of them based on speed and economy.

   There are other, less popular placement methods, such as belt conveyors and shotcrete.

Curing: *Curing* refers to the hydration process that helps maintain the chemical reaction that brings about concrete hardening and attaining strength. Curing can be a critical task in hot and dry locations. Wet pieces of fabric, such as canvas or burlap, may be used to cover the concrete from direct sunlight and keep moisture constantly in contact with the freshly placed concrete. Curing concrete in cold areas may require the addition of special admixtures and/or special heaters.

Concrete Finishing: Some formwork materials in direct contact with concrete may have chips and cracks or other imperfections on their surfaces that can be transferred to the concrete. If the concrete surface is visible, it then requires finishing by applying grout (or slurry) with a trowel or hand trowel. Finishing can also add texture or color to the concrete surface. Embossing (stamping) and coloring concrete, for example, can give it the appearance of masonry, brick, or even wood grain surface. However, many visible concrete surfaces are done without finishing by using smooth-surface forms and careful stripping tactics. This approach can produce significant savings.

Estimating Concrete

The *BCCD* contains two estimating methods: a detailed method and a shortcut. In the detailed method, at least four items must be estimated separately: formwork, reinforcement, concrete mix, and concrete placement. In the shortcut method, these four tasks have been combined and the data are shown as a mini-assembly on a single line.

Formwork Estimating

Before beginning a formwork estimate, the estimator must identify:

- The type of concrete member (e.g., suspended slab, column, etc.) along with any other criteria that may apply (e.g., height or size)
- Type of formwork material: wood, metal, fiber, and so on
- Number of uses (for wood forms)
- Quantity and unit
- Additional cost items, such as keyway for footing, or extra cost for height (suspended slabs or beams), battered walls, and so on
- Ceiling height, especially when it is not typical

Example 1

Estimate the total bare cost and the cost per SFCA for furnishing and erecting forms in place for 40, 16″ × 16″ square columns 10′ high, using plywood forms. Construction will be in Nashville, Tennessee, and forms will be used twice. Also estimate the total cost including overhead and profit (O&P) (Figure 6.1).

Solution

Use RSMeans line item 03 11 13.25 6050, adjusted for Nashville, TN:

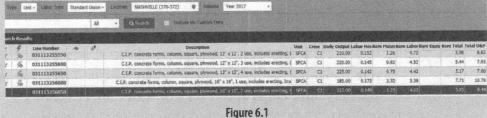

		Line Number			Description	Unit	Crew	Daily Output	Labor Hou	Bare Mater	Bare Labor	Bare Equip	Bare Total	Total O&P
r	%	031113255550			C.I.P. concrete forms, column, square, plywood, 12″ x 12″, 2 use, includes erecting, t	SFCA	C1	210.00	0.152	1.26	4.72		5.98	8.65
r	%	031113255800			C.I.P. concrete forms, column, square, plywood, 12″ x 12″, 3 use, includes erecting, t	SFCA	C1	220.00	0.145	0.92	4.52		5.44	7.93
r	%	031113255650			C.I.P. concrete forms, column, square, plywood, 12″ x 12″, 4 use, includes erecting, t	SFCA	C1	225.00	0.142	0.75	4.42		5.17	7.60
r	%	031113256000			C.I.P. concrete forms, column, square, plywood, 16″ x 16″, 1 use, includes erecting, bra	SFCA	C1	185.00	0.173	2.32	5.39		7.71	10.79
		031113256050			C.I.P. concrete forms, column, square, plywood, 16″ x 16″, 2 use, includes erecting, t	SFCA	C1	215.00	0.149	1.23	4.62		5.85	8.44

Figure 6.1

Formwork area = No. columns × 4 sides each × Width × Height
= 40 × 4 × 16″/12 in./ft. × 10′ = 2,133 SFCA

or

Unit bare cost = \$5.85/SFCA
Total bare cost for formwork = 2,133 SFCA × \$5.85/SFCA = \$12,478

The daily output of the crew (C1) in the previous item is 215 SFCA/day. The duration of this activity can be calculated as follows:

Duration = 2,133 SFCA/215 SFCA/day = 9.92 days ≈ 10 days

Example 2

Find the total cost including overhead and profit for the column formwork in Example 1 using:

a. RSMeans total incl. O&P.
b. RSMeans bare cost plus labor add-ons and 10 percent of the total direct costs for overhead and profit.

Solution

a. Total cost including O&P = Total area SFCA × Unit cost

= 2,133 SFCA × $8.44 / SFCA = $18,003 ≈ $18,000

b. From R013113–60, workers' compensation (WC) insurance for carpenters is 13.9 percent (average, with range of 4.7 to 39.7 percent) and Tennessee has a factor of 81 (i.e., 81 percent of the national average). Thus WC insurance in Tennessee is 13.9 × 0.81 = 11.26 percent. From *Installing Contractor's Overhead and Profit* (from the *BCCD* inside back cover and RSMeans Online estimating, Reference Items, Labor Rates), average fixed overhead is 18.3 percent. Overhead is 11 percent for carpenters, and profit is 10 percent. Total labor add-ons are 50.56 percent.

Unit prices for materials and labor next are adjusted for location from Example 1:

Bare cost for materials	= $1.23 × 2,133	= $ 2,624
Profit on materials	= $2,624 × 10%	= $ 262
Bare cost for labor	= $4.62 × 2,133	= $ 9,854
Add-ons and profit for labor	= $ 9,854 × 50.56%	= $ 4,982
		$17,722
Total		≈ $17,700

Note that the two results are slightly different (by $281, or about 1.5 percent). Since cost estimating is not an exact science, each approach uses different assumptions, resulting in answers that may vary. If those assumptions were reasonable and compatible with the considered situation, differences among answers should be minimal.

Using Local Labor Wages: In the next example, substitute local wages for the RSMeans labor union wages. It is not necessary to apply a location factor adjustment to the local labor cost.

Example 3

Repeat Example 1 with given local labor rates of $35/hour for carpenters and $29/hour for laborers.

Solution

Crew C1 is composed of three carpenters and one laborer. The adjusted daily cost for crew C1 is:

$$3 \text{ carpenters:} \quad 3 \times \$35/\text{hour} \times 8 \text{ hours}/\text{day} = \$\ \ 840.00$$
$$1 \text{ laborer:} \quad\quad 1 \times \$29/\text{hour} \times 8 \text{ hours}/\text{day} = \$\ \ 232.00$$
$$\text{Total bare cost per day for local crew C1} \quad = \$1,072.00$$

Since the crew's daily productivity is 215 SFCA/day, the labor cost per SFCA is $1,072/215 = $4.99/SFCA.

Material cost (from Example 1) = $1.23/SFCA

Total bare cost = $1.23 + $4.99 = $6.22/SFCA

Total bare cost for formwork = 2,133 SFCA × $6.22/SFCA = $13,267

For calculating the cost including O&P, follow the same procedure as Example 2:

Bare cost for materials (from Example 1)	= $1.23/SFCA × 2,133 SFCA	= $ 2,624
Profit on materials	= $2,624 × 10%	= $ 262
Bare cost for labor	= $4.99/SFCA × 2,133 SFCA	= $10,644
Add-ons and profit for labor	= $10,644 × 50.56%	= $ 5,381
Total		= $18,911
		≈ $18,900

Adjusting Productivity Rates: Job productivity may differ from that used in the *BCCD* for a variety of reasons, such as complexity of the job, weather conditions, or overtime. In the case of overtime, both productivity and wages must be adjusted. In all cases, productivity has to be adjusted with regard to labor and equipment costs. For more details see the section titled "Overtime Productivity Loss and Extra Pay" in Chapter 5.

Example 4

Repeat Example 1 with the condition that the task involved will take place during a hot and humid summer, which tends to cut productivity by about 25 percent (Figure 6.2).

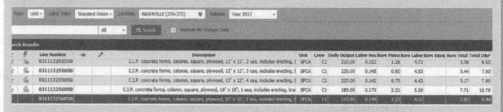

		Line Number			Description	Unit	Crew	Daily Output	Labor Hou	Bare Mate	Bare Labor	Bare Equi	Bare Total	Total O&P
		031113255550			C.I.P. concrete forms, column, square, plywood, 12" x 12", 2 use, includes erecting,	SFCA	C1	210.00	0.152	1.26	4.72		5.98	8.65
		031113255600			C.I.P. concrete forms, column, square, plywood, 12" x 12", 3 use, includes erecting,	SFCA	C1	220.00	0.145	0.92	4.52		5.44	7.93
		031113255650			C.I.P. concrete forms, column, square, plywood, 12" x 12", 4 use, includes erecting,	SFCA	C1	225.00	0.142	0.75	4.42		5.17	7.60
		031113256000			C.I.P. concrete forms, column, square, plywood, 16" x 16", 1 use, includes erecting, bra	SFCA	C1	185.00	0.173	2.32	5.39		7.71	10.79
		031113256050			C.I.P. concrete forms, column, square, plywood, 16" x 16", 2 use, includes erecting,	SFCA	C1	215.00	0.149	1.23	4.62		5.85	8.44

Figure 6.2

Solution

$$\text{Daily output} = 215\,\text{SFCA} \times (100\% - 25\%) = 161\,\text{SFCA}$$
$$\text{Labor hours (per SFCA)} = 0.149/(100\% - 25\%) = 0.199$$

or

Crew labor hours / Daily output = Labor hours / unit
32 LH / 161 SFCA per day = 0.199 LH per SFCA
(Where 32 is total labor-hours per day for Crew C1)

$$\text{Unit bare cost} = \$1.23 + \$4.62/(100\% - 25\%) = \$7.39$$
$$\text{Total bare cost for formwork} = 2{,}133\,\text{SFCA} \times \$7.39/\text{SFCA} = \$15{,}763$$

Materials are not subject to productivity adjustments. If equipment was used in this item, its cost has to be adjusted for the change in productivity in the same manner as labor. Equipment usage duration is equal to the labor duration.

Example 5

Find the total cost including O&P for the formwork in Example 4.

Solution

We have to separate labor from material cost in the *BCCD*.
The material bare cost is $1.23/SFCA (adjusted for Nashville, Tennessee).

There is an assumed 10 percent profit on materials.

Thus, materials unit cost + profit = $1.23 \times (100\% + 10\%) = \$1.35/\text{SFCA}$[1]

Labor add-ons = 50.56% (see Example 2)

Then labor unit cost incl. O&P (before productivity adjustment)

$$= \$4.62 \times (100\% + 50.56\%) = \$6.96/\text{SFCA}$$

Unit cost incl. O&P = $\$1.35 + [\$6.96/(100\% - 25\%)] = \$10.63$

Total cost incl. O&P = 2,133 SFCA $\times \$10.63/\text{SFCA} = \$22,674 \approx \$22,700$

Example 6

Repeat Example 1 with the crew working 10 hours per day, 6 days per week. Overtime hours are compensated at 1.5 times the regular pay.

Solution

Assume the job duration is two weeks (see Example 1). Use reference number R012909–90, as shown in Figure 6.3.

General Requirements		R0129 Payment Procedures					

R012909-90 Overtime

One way to improve the completion date of a project or eliminate negative float from a schedule is to compress activity duration times. This can be achieved by increasing the crew size or working overtime with the proposed crew.

To determine the costs of working overtime to compress activity duration times, consider the following examples. Below is an overtime efficiency and cost chart based on a five, six, or seven day week with an eight through twelve hour day. Payroll percentage increases for time and one half and double time are shown for the various working days.

Days per Week	Hours per Day	Production Efficiency					Payroll Cost Factors	
		1st Week	2nd Week	3rd Week	4th Week	Average 4 Weeks	@ 1-1/2 Times	@ 2 Times
	8	100%	100%	100%	100%	100 %	100 %	100 %
	9	100	100	95	90	96.25	105.6	111.1
5	10	100	95	90	85	91.25	110.0	120.0
	11	95	90	75	65	81.25	113.6	127.3
	12	90	85	70	60	76.25	116.7	133.3
	8	100	100	95	90	96.25	108.3	116.7
	9	100	95	90	85	92.50	113.0	125.9
6	10	95	90	85	80	87.50	116.7	133.3
	11	95	85	70	65	78.75	119.7	139.4
	12	90	80	65	60	73.75	122.2	144.4
	8	100	95	85	75	88.75	114.3	128.6
	9	95	90	80	70	83.75	118.3	136.5
7	10	90	85	75	65	78.75	121.4	142.9
	11	85	80	65	60	72.50	124.0	148.1
	12	85	75	60	55	68.75	126.2	152.4

Figure 6.3

Production efficiency (average two weeks) = (95% + 90%) = 92.5%

Payroll cost factor = 116.7%

Unit bare cost = $\$1.23 + [(\$4.62/92.5\%) \times 116.7\%]$ = $7.06

Total bare cost for formwork = 2,133 SFCA $\times \$7.06/\text{SFCA}$ = $15,056

≈ $15,100

---

1. In carrying out such calculations, it is better not to round early (at the unit level). We are showing the numbers rounded here for simplicity, but only the final answer should be rounded.

In order to calculate duration, we need to find the daily output:

Daily output = 215 SFCA /day × 92.5% × 10 / 8 day = 249 SFCA /day
Duration = 2,133 SFCA / 249 SFCA /day = 8.6 days

Duration is then 8½ days as compared to 10 regular days. But bear in mind that the original 10 days are equivalent to 2 weeks (5 workdays per week), while the overtime 8½ days are about 1½ weeks (with 6 workdays per week).

Using Forms More than Four Times: Most items in the concrete formwork section (wood and plywood forms) of RSMeans cost data are based on one, two, three, or four uses. Formwork cost is a combination of materials and labor cost. The labor cost includes making up, erecting, stripping, cleaning, and moving. The materials cost includes the lumber, plywood, and other accessories needed, plus an allowance for each reuse. See R031113–40 for details. In cases of more than four uses of the forms, we have to calculate the cost of the average use.

Example 7

Calculate the total bare cost per SFCA for furnishing and erecting plywood forms for 16 columns, 24″ × 24″ in cross-section and 12' tall. Forms will be used five times, and the job is in Roswell, New Mexico.

Solution

Total quantity = 16 × (24″/12) × 12' × 4 sides = 1,536 SFCA

Examine line items 03 11 13.25 6500–6650 (keep it in National Average cost).

From R031113–40, the material cost for reusing columns includes 10 percent replacement (waste) for each reuse.

For the first use, from line 03 11 13.25 6500, the bare material cost per SFCA = $2.70.

For the second use, add 10 percent for repair/replacement:

Bare material cost per SFCA = ($2.70 × 1.1)/2 = $1.49

For the third use:

Bare material cost per SFCA = ($2.71 × 1.2)/3 = $1.08

For the fourth use:

$$\text{Bare material cost per SFCA} = (\$2.71 \times 1.3)/4 = \$0.88$$

For the fifth use:

$$\text{Bare material cost per SFCA} = (\$2.71 \times 1.4)/5 = \$0.76$$

To calculate the labor cost, we first must determine the labor hours required per use. From R031113–40, the labor cost for one hour of carpenter time for repairs for each reuse (per 100 SFCA) should be included. Using this logic, the labor hours for formwork in R031113–60 can be used to determine the labor cost for each reuse, as shown next. For the first use, the labor required is 5.8 LH to fabricate, 9.8 LH to erect and strip, and 1.2 LH to clean and move the forms.

$$5.8\,\text{LH} + 9.8\,\text{LH} + 1.2\,\text{LH} = 16.8\,\text{LH}$$

For each reuse, the LH required are 9.8 LH to erect and strip, 1.2 LH to clean and move, and 1 LH to repair.

$$9.8\,\text{LH} + 1.2\,\text{LH} + 1\text{LH} = 12\,\text{LH}$$

The labor hours required, then, to use the forms twice is:

$$16.8\,\text{LH} + 12\,\text{LH} = 28.8\,\text{LH}$$

and the average labor hours per use is:

$$(16.8\,\text{LH} + 12\,\text{LH})/2 = 14.4\,\text{LH}$$

Using the same logic:
For three uses:

$$(16.8\,\text{LH} + 12\,\text{LH} + 12\,\text{LH})/3 = 13.6\,\text{LH}$$

For four uses:

$$(16.8\,\text{LH} + 12\,\text{LH} + 12\,\text{LH} + 12\,\text{LH})/4 = 13.2\,\text{LH}$$

For five uses:

$$(16.8\,\text{LH} + 12\,\text{LH} + 12\,\text{LH} + 12\,\text{LH} + 12\,\text{LH})/5 = 13\,\text{LH}$$

The average hourly bare cost for labor in Crew C1 is $((49.25 \times 3 + 39.15) / 4) = \$46.73$. Multiplying the average labor hours required by the cost per hour will give the average labor cost for forms used five times:

$$13\,\text{LH} \times \$46.73/\text{LH} = \$607.43 \text{ per 100 SFCA, or } \$6.07 \text{ per SF.}$$

CCI for Roswell, New Mexico, Division 03100: 99.5%, 65.7%, 70.3% for materials, installation, and total, respectively.

Average unit material cost for five uses:

$0.76/ SFCA × 99.5% = $0.76/ SFCA

Average unit labor cost for five uses:

$6.07/SFCA × 65.7% = $3.99/SFCA

Average total bare cost = $0.76 + $3.99 = $4.75/ SFCA

Total bare cost = 1,536 × $4.75 = $7,296 ≈ $7,300

If the item is not included in R031113–60, use the same concept to derive the costs from the *BCCD* item rather than the Reference section. See Example 8 for details.

Example 8

Calculate the total bare cost per SFCA for furnishing and erecting forms for a continuous footing, 650′ long, 3′ wide, and 16″ high. Forms will be used five times, and the job is in Roswell, New Mexico (Figure 6.4).

Solution

Total formwork area = (650′ × 16″/12 × 2 sides) + (3′ × 16″/12 × 2)
= 1,741 SFCA

Lines 03 11 13.45 0010–0150 show the unit cost for the item for uses 1 through 4.

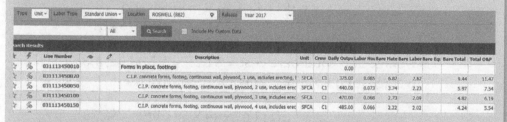

Figure 6.4

Assume that the cost of the first use is *X* and the cost of each reuse is *Y*.

The average cost per use in case of two uses = $(X + Y)/2$
for three uses = $(X + 2Y)/3$
for four uses = $(X + 3Y)/4$, and so on.

From item 03 11 13.45 0020 (one use), material cost only, we find that $X = \$6.82$.

From item 03 11 13.45 0050 (two uses), $(X + Y) / 2 = \$3.74$, which is the average of first use and one reuse.
Then:

$$\text{Let } X = \text{the cost of the first use}$$
$$Y = \text{the cost of each reuse}$$
$$X = \$6.82 \text{ and } (X + Y)/2 = \$3.74$$

Then, solving for Y:

$$(\$6.82 + Y)/2 = \$3.74$$
$$Y = (2 \times \$3.74) - \$6.82 = \$0.66$$

Average cost of materials with five uses:

$$[\$6.82 + (4 \times \$0.66)]/5 = \$1.89$$

For labor cost:

$$X = \$2.62, (X + Y)/2 = \$2.23, \text{ then}$$
$$(\$2.62 + Y)/2 = \$2.23, \text{ and}$$
$$Y = (2 \times \$2.23) - \$2.62 = \$1.84$$

Average cost of labor for five uses:

$$[\$2.62 + (4 \times \$1.84)]/5 = \$2.00$$

For total cost including O&P:

$$X = \$11.47, (X + Y)/2 = \$7.54, \text{ then}$$
$$(\$11.47 + Y)/2 = \$7.54, \text{ and}$$
$$Y = (2 \times \$7.54) - \$11.47 = \$3.61$$

Average unit total bare cost = \$1.89 + \$2.00 = \$3.89 / SFCA
Average total cost including O&P for five uses:

$$[\$11.47 + (4 \times \$3.61)]/5 = \$5.18$$

$$\text{Total bare cost} = 1,741\,\text{SFCA} \times \$3.89/\text{SFCA} = \$6,772$$
$$\text{Total cost incl. O\&P} = 1,741\,\text{SFCA} \times \$5.18/\text{SFCA} = \$9,018$$

We can solve this example based on the cost of one and four uses:

$$X = \$6.82 \text{ and}$$
$$\frac{X + 3Y}{4} = \$2.22, \text{ then}$$

$$\frac{\$6.82 + 3Y}{4} = \$2.22, \text{ and}$$

$$Y = \frac{4 \times \$2.22 - \$6.82}{3} = \$0.69$$

Average cost of materials with five uses $= \dfrac{\$6.82 + (4 \times \$0.69)}{5} = \$1.91,$

which is close to the result we obtained earlier. Difference is caused by rounding.

Estimating Reinforcement

Reinforcement is usually estimated in units of weight, such as tons. Generally, labor cost per ton increases as bar size decreases since there are more bars/ton for the smaller size to handle than the larger size. Some estimators like to estimate reinforcement in linear feet. In the case of welded wire fabric, it is estimated in area units, mostly in CSF, which is 100 square feet (C in Roman numerals = 100). Remember to include allowances for splices, overlaps, and waste.

For epoxy-coated or galvanized reinforcement, you have to add extra cost. There are also special types of reinforcement bars, such as glass fiber-reinforced polymer bars, which are priced differently. Also, synthetic fiber reinforcing may be added to the concrete mix as a substitute for reinforcement in some slabs on grade or other members subject to relatively low stresses. New materials may be introduced to the market that may not be listed in the database. All these items have to be carefully priced.

Example 9

Estimate the total bare cost for furnishing and erecting reinforcement for the 40 columns in Example 1. Eight #10 bars will be used for each column, with a total length of 13′ for each bar. Stirrups will be of #4 at 12″ spacing. All steel is Grade 60, and construction will be in Nashville, Tennessee.

Solution
Longitudinal (main) reinforcement:

Total number of bars	= 40 columns × 8 bars/column	= 320 bars
Total length of bars	= 320 bars × 13′ each	= 4,160 LF

Actual length should be estimated from the plans including any required lap lengths.

R032110–10 gives weight per LF for rebar. For #10, it is 4.303 lb/LF.

$$\begin{aligned}
\text{Total weight of bars} &= 4{,}160\,\text{LF} \times 4.303\,\text{lb/LF} \\
&= 17{,}900\,\text{lb} \\
&= 17{,}900\,\text{lb}/2{,}000\,\text{lb/ton} \\
&= 8.95\,\text{tons}
\end{aligned}$$

Use RSMeans line item 03 21 11.60 0250 with location adjusted to Nashville, TN:

RSMeans Item No.	Description	Crew	Daily Output	Bare Costs				Total Incl. O&P
				Mat.	Labor	Equip.	Total	
03 21 11.60 0250	Reinf. in place, Gr. 60, column #8-#18	4 Rdmn	2.3 tons	898.64	487.73	–	1,386.37	1,738.95

$$\begin{aligned}
\text{Total bare cost} &= 8.95\,\text{tons} \times \$1{,}386.37/\text{ton} = \$12{,}408 \\
\text{Total cost incl. O\&P} &= 8.95\,\text{tons} \times \$1{,}738.95/\text{ton} = \$15{,}564
\end{aligned}$$

Stirrups

The length of a stirrup has to be calculated according to the design. In this example, it is assumed that the stirrup length is:

$$13'' \text{ for each side} \times 4 \text{ sides} + 4.5'' \times 2 \text{ hooks} = 61'' = 5.083'$$

Total number of stirrups is 11 per column.

$$\begin{aligned}
\text{Total length of stirrups} &= 40 \text{ columns} \times 11 \text{ stirrups/column} \times 5.083' \\
&= 2{,}237\,\text{LF}
\end{aligned}$$

From R032110–10, we find the weight per LF for #4 rebar is 0.668 lb/LF.

$$\text{Total weight of stirrups} = 2{,}237\,\text{LF} \times 0.668\,\text{lb/ft} = 1{,}494\,\text{lb} = 0.747\,\text{tons}$$

Use RSMeans item number 03 21 11.60 0200:

RSMeans Item No.	Description	Crew	Daily Output	Bare Costs				Total Incl. O&P
				Mat.	Labor	Equip	Total	
03 21 10.60 0200	Reinf. in place, Gr. 60, columns #3 to #7	4 Rdmn	1.5 tons	898.64	742.90	–	1,641.54	2,126.55

Total bare cost	= 0.747 tons × $1,641.54/ton	= $1,226
Total cost incl. O&P	= 0.747 tons × $2,126.55/ton	= $1,589
Total bare cost for reinforcement	= $12,408 + $1,226	= $13,634
Total cost for reinforcement incl.O&P	= $15,564 + $1,589	= $17,153

Note: We did not round up the final answers here (except for fractions of a dollar) because, most likely, these numbers are part of a bigger estimate, so only the final numbers are to be rounded.

Example 10

Calculate the cost for furnishing and installing welded wire fabric (WWF) size 6 × 6 – W2.9 × W2.9 for a 120′ × 75′ slab on grade. The job will be in San Francisco, California.

Solution

Total slab area = 120′ × 75′ = 9,000 SF

Welded wire fabric is usually supplied in sheets. Assume a 10 percent waste factor for overlaps.

Total quantity for WWF = 9,000 SF (1 + 10%) = 9,900 SF = 99 CSF

Use RSMeans line item 03 22 11.10 0300, adjusted for San Francisco, California:

RSMeans Item No.	Description	Crew	Daily Output	Bare Costs				Total Incl. O&P
				Mat.	Labor	Equip.	Total	
03 2211.10 0300	WW fabric, 6 × 6 - W2.9 × W2.9	2 Rdmn	29 CSF	25.63	38.37	–	64.00	87.07

Total bare cost = 99 *CSF* × $64.00 = $6,336
Total cost incl. O&P = 99 *CSF* × $87.07 = $8,620

Estimating Concrete Mix and Placement

The concrete mix has to be chosen according to the required strength and whether it is field mix or ready mix. Admixtures or coloring pigments add extra costs. The cost of placement must also be included in an estimate.

Example 11

Estimate the bare cost of furnishing and placing concrete for the columns in Example 1.[2] Assume a column height of 10′. Concrete will be pumped and will have a minimum compressive strength of 3,500 pounds per square inch (psi).

Solution

Volume of concrete per column $=$ Width $\times$ Thickness $\times$ Length
$= 16''/12 \times 16''/12 \times 10' = 17.78$ CF
$= 17.78$ CF/27 $= 0.66$ CY

Total concrete volume $= 40$ columns $\times 0.66$ CY/column $= 26.34$ CY

Assume a 5 percent waste factor that applies to materials cost only.

RSMeans item No.	Description	Crew	Daily Output	Bare Costs				Total Incl. O&P
				Mat.	Labor	Equip	Total	
03 31 13.35 0200	Concrete ready mix 3500 psi	–	–	110.11	–	–	110.11	121.03
03 31 13.70 0600	Placing concrete columns square 18″ thick pumped	C20	90 CY	–	19.98	9.93	29.91	41.21

The columns being used are 16 inches thick. The only lines in the cost database, however, are for 12-inch- and 18-inch-thick columns. Use the closest item—that is, the 18-inch column. Interpolation between the two items is also possible.

Bare cost for mix	$=$ Volume $\times$ (1 + Wastefactor) $\times$ Unit cost	
Bare cost for mix	$= 26.34$ CY $\times 1.05 \times \$110.11$/CY	$= \$3,045$
Bare cost for placement	$= 26.34$ CY $\times \$29.91$	$= \$\ \ \ 788$
Total bare cost	$= \$3,045 + \$788$	$= \$3,833$
Cost incl. O&P for mix	$= 26.34$ CY $\times 1.05 \times \$121.03$/CY	$= \$3,347$
Cost incl. O&P for placement	$= 26.34$ CY $\times \$41.21$	$= \$1,086$
Total cost incl. O&P	$= \$3,347 + \$1,086$	$= \$4,433$

Proportional Quantities: Use R033105–10 for calculating the formwork area and estimating the reinforcement as a function of concrete quantity. An important related note here is the change in the RSMeans *BCCD*

---

2. Note that the dimensions the structural engineers use may be different from those used by quantity/cost estimators. Basically, estimators deduct any shared amount, such as the top of the column, that has been already included in the quantity of the beam and/or suspended slab above the column.

categorization of reinforcement of columns. In past editions of the *BCCD*, columns in section 03 30 53.40 were listed as minimum, average, and maximum reinforcing. In the newer edition, this category was changed to "Up to 1%," "Up to 2%," and "Up to 3%" reinforcing by area.

Example 12

Calculate the amount of formwork for Example 1, and estimate the amount of rebar needed, assuming average reinforcement.

Solution

Concrete volume = 26.34 CY (from Example 11)

$$\text{Formwork} : \text{Using R033105} - 10 = 26.34 \text{ CY} \times 81 \text{ SFCA/CY} = 2{,}133 \text{ SFCA}$$

This is the same result we obtained in Example 1.
R033105–10 provides a range of figures for reinforcement of 16-inch square columns, from a maximum of 1,082 lb/CY to a minimum of 187 lb/CY. The arithmetic average is 635 lb/CY.

$$\text{Total weight of steel} = 26.34 \text{ CY} \times 635 \text{ lb/CY} = 16{,}726 \text{ lb} = 8.363 \text{ tons}$$

This is close to the 8.95 tons we calculated in Example 9 but since the variations in the amount of reinforcing steel could be large, using these tables may give relatively large error / difference.
Since we don't have the breakdown of the rebar quantity into main (longitudinal) bars and stirrups, and since the main bars make up the biggest percentage of the total weight, use the cost for the bars calculated in Example 9.
Total cost:

$$\text{Total adjusted bare cost} = 8.363 \text{ tons} \times \$1{,}641.54/\text{ton} = \$13{,}728$$
$$\text{Total cost including O\&P} = 8.363 \text{ tons} \times \$2{,}126.55/\text{ton} = \$17{,}784$$

Mini-Assemblies (Shortcut Method)

Division 3 of the *BCCD* has two methods of estimating concrete. The line item unit cost method was detailed earlier. The other method, a mini-assembly, requires only one item. However, in exchange for this convenience, one has to concede to the assumptions made for that mini-assembly. Users must read the description carefully and use the line item only if the description fits the needed item.

Example 13

Estimate the bare cost and cost including overhead and profit for building the 40 columns in Example 1 using a mini-assembly.

Solution

Use the shortcut method, and go to line number 03 30 53.40 0820:

RSMeans Item No.	Description	Crew	CY/Day	Bare Costs				Total Incl. O&P
				Mat.	Labor	Equip	Total	
03 30 53.40 0820	Concrete in place incl. forms reinf. Column 16" × 16" 2%–3% reinf	C14A	12.57	436.80	522.81	68.76	1,028.37	1,352.43

Concrete volume = 26.34 CY (from Example 11)

Total bare cost = 26.34 CY × $1,028.37   = $27,087
Total cost incl. O&P = 26.34 CY × $1,352.43 = $35,623

Comparing these results with the detailed method:

Bare Cost:
    Formwork (from Example 1) $12,478
    Reinforcement (from Example 9) $13,634
    Concrete mix and placement (from Example 11) $ 3,833
    Total bare cost $29,945
Cost Including O&P:
    Formwork (from Example 2) $18,003
    Reinforcement (from Example 9) $17,153
    Concrete mix and placement (from Example 11) $ 4,433
    Total cost incl. O&P $39,589

Extreme Weather Cost Factors

Concrete hardening is a chemical reaction that must take place within certain climatic conditions. When the weather turns cold, the reaction slows down, and the needed strength of the concrete may not be attained at the desired speed. In very cold temperatures, the needed strength may not be attained at all. Depending on the severity of the cold, special measures may have to be taken to ensure the strength of the concrete, including—but not limited to—adding accelerating agents to the mix, heating the concrete, insulating the formwork, and using electrically heated blankets. In extremely hot weather, curing becomes a critical activity, especially when the concrete member is under the direct rays of the sun. The chemical reaction that results in concrete hardening requires, among other things, a certain amount of moisture. Concrete curing may include items such as burlap blankets. This serves two purposes: providing moisture to the concrete and protecting it from the heat of the sun.

Example 14

Estimate the total bare cost of furnishing and installing concrete for 30 spread footings, all 4′ × 4′ × 2′. Concrete should have a minimum compressive strength of 4,000 psi. Construction will be in Minneapolis, Minnesota, during the winter season. Electrically heated pads with 15 watts/SF will be used. Concrete will be placed by direct chute.

Solution

$$\text{Concrete volume per footing} = (\text{Length}' \times \text{Width}' \times \text{Depth}')/27 \text{ CF/CY}$$
$$= (4' \times 4' \times 2')/27 = 1.185 \text{ CY}$$
$$\text{Total concrete volume} = 30 \text{ footings} \times 1.185 \text{ CY each} = 35.6 \text{ CY}$$

Use the following RSMeans line items, already adjusted to Minneapolis, Minnesota:

RSMeans Item No.	Description	Daily Output	Unit	Bare Costs				Total Incl. O&P
				Mat.	Labor	Equip	Total	
03 31 13.35 0300	Conc. ready mix 4,000 psi	–	CY	126.88	–	–	126.88	139.06
03 31 13.35 1300	Struct conc., winter mix (hot water), add	–	CY	4.52	–	–	4.52	4.97
03 31 13.70 2400	Place conc. ftng. < 1CY chute	55	CY	–	40.04	1.11	41.15	62.13
03 05 13.85 0710	Winter protec, electr heated pads, 110 volts, 15 watts per SF, 20 uses	–	SF	.55	–	–	.55	.60

A few assumptions have been made here:

1. We will use the item from spread footing for less than 1 CY. Our footing volume is just over 1 CY. One can interpolate between the items, less than 1 CY, and over 5 CY.

2. The contractor will rent the electrically heated pads. We interpolated the above item as an average between the minimum and the maximum.

3. Assume the area to be heated is the sides and top, that is:
    30 footings × 4 sides each × 4' × 2' + (4' × 4') top = 1,440 SF.

4. Assume a $0.18/SF labor allowance for the installation of the heating pads. If you look at burlaps, item no. 03 39 13.50 0015, you'll see that the labor cost is $12.86/CSF ≈ $0.13/SF. Since it takes a little more time to install electrical pads than burlaps, assume bare cost of $0.18/SF and $0.27/SF including O&P.

5. No productivity adjustment was done. Realistically, productivity decreases in cold weather.[1]

Bare cost:

$$\text{Concrete mix} = 35.60 \text{ CY} \times 1.05 \text{ (waste factor)} \times (\$126.88 + 4.52)/\text{CY} = \$4{,}912$$

Placing concrete = 35.60 CY × $41.15/CY = $ 1,465
Electrically heated pads = 1,440 SF × $0.55/SF = $ 792
Installation of heated pads = 1,440 SF × $0.18/SF = $ 259
Total bare cost = $7,428

Total cost including O&P:

Concrete mix = 35.60 CY × 1.05 × ($139.06 + 4.97)/CY = $ 5,384
Placing concrete = 35.60 CY × $62.13/CY = $ 2,212
Electrically heated pads = 1,440 SF × $0.60/SF = $ 864
Installation of heated pads = 1,440 SF × $0.27/SF = $ 389
Total cost incl. O&P = $8,849

Estimating Precast Concrete

Precast concrete is available in two types:

1. *Structural*, which includes members such as beams, slabs, Ts and double Ts, stairs, and columns

2. *Architectural*, which includes precast panels and other non-load-bearing members

Some companies specialize in precast concrete, and sometimes in certain types of precast. Most precast concrete manufacturers include delivery and sometimes installation with the material cost.

Since the equipment portion is very small in regard to precast concrete, we used the CCI "Inst." factor for placement (labor and equipment). The term

---

1. The author has several studies on optimum scheduling, which can be summarized as selecting the project's starting point and the composition of the durations and timing of its activities, within logic constraints, that will result in optimum schedule and least cost while maintaining the project's scope and quality. See *Construction Project Scheduling and Control* by Saleh Mubarak, 4th ed., 2019.

non-load-bearing in the structural design context means that the member is not designed to carry a structural load even though it may carry some load.

Example 15

Estimate the total bare cost for furnishing and erecting a precast concrete double T system to cover three suspended floors. Members to be used are (per floor):

1. 20 double Ts, standard weight, 18″ × 8′ wide, 30′ long

2. 9 columns, 24″ × 24″, 12′ high

Construction is to take place in Los Angeles, California.

Solution

The quantities to be used are:

Double Ts = 20 per floor × 3 floors = 60 Ea

Columns = 9 per floor × 12′ high × 3 floors = 324 LF

RSMeans Item No.	Description	Daily Output	Unit	Bare Costs				Total Incl. O&P
				Mat.	Labor	Equip	Total	
03 41 33.60 1350	Double Ts, std wt. 18″× 8′ wide, 30′ span	20	Ea	2,470.65	208.36	101.30	2,780.31	3,191.89
03 41 33.15 0050	Column, rectangular, to 12′ high, 24″ x 24″	96	LF	313.85	43.50	21.27	378.62	436.27

Bare cost = 60 double T × $2,780.31 + 324 LF columns × $436.27

= $166,819 + 141,351 = $308,170

≈ $308,000

Cost incl. O&P = 60 double T × $3,191.89 + 324 LF columns × 436.27

= $191,513 + 141,351 = $332,864

≈ $333,000

When comparing precast concrete to cast-in-place (CIP) concrete, it is interesting to notice the high cost for materials as compared to installation (labor and equipment). The material cost for precast actually includes all the labor and equipment needed for the manufacturing of the members. Labor and equipment cost shown here includes field installation only. Obviously, a precast unit takes a lot less time and effort to install than a similar CIP member.[2]

---

2. The making of precast units, including 3-D modules, is considered manufacturing, not construction. In construction projects, there are always some manufactured elements, small or large, that are influenced by feasibility, economy, and schedule urgency. Refer to Mubarak, *Construction Project Scheduling and Control*, 4th edition, chapter 8, "Schedule Compression and Time–Cost Trade-Off," Construction and Modularization, pp. 235-236.

Lightweight Concrete

Lightweight concrete basically includes two types: structural lightweight concrete and insulating concrete. Regular concrete usually weighs about 150 PCF (pounds per cubic foot). Lightweight concrete weighs between 20 and 120 PCF. Structural lightweight concrete is usually in the upper range of that weight, while insulating concrete can be as light as 20 PCF or less.

Reducing the weight of concrete is usually achieved in two steps:

1. Replace all or a portion of the aggregates and/or sand by other lighter materials, such as fly ash, slag, perlite, vermiculite, volcanic ash, fiber (polypropylene or other), or foam.

2. Aerate (air-entrain) the concrete to form millions of tiny air bubbles. Aerated lightweight concrete is made under high pressure to force the air inside the mix.

Lightweight concrete is used as a concrete mix for special situations and premanufactured members, such as deck planks, steps, and block.

Example 16

An office building in downtown Chicago, Illinois, with a flat roof needs to be insulated with 3″ lightweight concrete. The building has a rectangular cross-section with dimensions of 85′ × 52′. Estimate the cost of providing and installing the insulating concrete.

Solution

The total area to be covered = 85′ × 52′ = 4,420 SF. Use the following RSMeans item:

RSMeans Item No.	Description	Unit	Bare Costs				Total Incl. O&P
			Mat.	Labor	Equip.	Total	
03 52 16.16 0250	Insulating roof fill, perlite/vermiculite, 1:6 ready mix, 3″ thick	SF	2.30	0.50	0.10	2.90	3.39

$$
\begin{aligned}
\text{Total bare cost} &= 4,420\,\text{SF} \times \$2.90 \\
&= \$12,818 \approx \$12,800 \\
\text{Total cost incl. O\&P} &= 4,420\,\text{SF} \times \$3.39 \\
&= \$14,984 \approx \$15,000
\end{aligned}
$$

Additional Estimating Examples

Example 17

Estimate the cost of the job in Example 1 if it will be done during summer 2021.

Solution

Since RSMeans prices are for January 2017, there are about 4½ years between the price quotes and the actual construction. Assume an average inflation of 4 percent for the next few years.

$$\text{Future bare cost} = \text{Present bare cost} \times (1 + i\%)^n$$

where i = expected annual inflation and n = the time interval between current prices time and midpoint of the studied project.

$$\text{Future bare cost} = \$12,478 \text{ (from Example 1)} \times (1.04)^{4.5} = \$14,887 \approx \$14,900$$

Example 18

For the columns in Example 11, assume that the equipment will be:

1. Rented and used for 2 days only.

2. Rented for a month and used for 18 days during the month, 8 hours/day.

Location: Nashville, Tennessee

Solution

Crew C20 was used in Example 11. This crew uses a small concrete pump that matches the line item 01 54 33.10 2120, and two gas engine vibrators that match the line item 01 54 33.10 3000.

1. The rental rate for the pump is $997.81, $2,998.60, and $8,995.80, for daily, weekly, and monthly rental periods.

2. The operating cost is $29.26/hour.

3. The crew equipment cost/day is $833.80 (explained in Chapter 1).

4. The rental rate for each vibrator is $15.87, $47.56, and $142.69, for daily, weekly, and monthly rental periods.

5. The operating cost is $2.46/hour.

6. The crew equipment cost/day is $29.19.

The cost based on two days' rental only:
Pump cost:

Daily rental = $997.81
Hourly operating cost = $29.26
Crew equip, cost = $997.81 + ($29.26/hr × 8 hrs/day) = $1,232/day

Vibrators cost:

Daily rental = $15.87
Hourly operating cost = $2.46
Crew equip, cost = $15.87 + ($2.46/hour × 8 hours/day)
= 35.55/day

Total crew equipment cost/day = One pump + Two vibrators
= $1,232 + (2 × $35.55) = $1,303

Total daily production (from Example 11) = 90 CY
Equipment cost per CY = $1,303/90 = $14.48/CY

Replace the equipment unit cost in Example 11 by $14.48/CY.

Bare cost for mix (same as Example 11) = $ 3,045
Bare cost for labor = 26.34 CY × $19.98 = $526.27
Bare cost for equipment = 26.34 CY × $14.48 = $381.40
Total bare cost = $ 3,953

The cost based on monthly rental and 18 days use during the month:
Pump cost:

Monthly rental = $8,995.80
Hourly operating cost = $29.26
Crew equip. cost = $8,995.80/18 + ($29.26/hr × 8 hrs/day)
= $733.85/day

Vibrators cost (each):

Monthly rental = $142.69
Hourly operating cost = $2.46
Crew equip. cost = $142.69/18 + ($2.46/hr × 8 hrs/day)
= $27.61/day

Total crew equipment cost/day = One pump + Two vibrators
= $733.85 + (2 × $27.61)
= $789

Total daily production (from Example 11) = 90 CY
Equipment cost per CY = $789/90 = $8.77/CY

Replace the equipment unit cost in Example 11 by $8.77/CY.

Bare cost for mix (same as Example 11)	= $ 3,045
Bare cost for labor(same as above)	= $526.27
Bare cost for equipment = 26.34 CY × $8.77/CY	= $ 231
Total bare cost	= $ 3,802

Comparing results: We find, from both Examples 11 and 18, that the bare cost for placing the concrete for the columns is $3,953, $3,833, and $3,802 for the cases of renting the equipment on a daily, weekly, and monthly basis, respectively. The differences are small here because the equipment cost is small compared to cost of materials and labor.

Example 19

Estimate the bare cost and the cost including O&P for providing and installing post-tension cables for six CIP suspended slabs. Each is 120′ long by 80″ wide and 9″ thick. Use an average of 1 lb/SF. Strands are grouted and will be pulled by a 200 kips force. The job will be in St. Petersburg, Florida.

Solution

Interpolate between *BCCD* line items 03 23 05.50 0300 and 0350.

RSMeans Item No.	Description	Daily Output	Unit	Bare Costs				Total Incl. O&P
				Mat.	Labor	Equip	Total	
03 23 05.50 0300	Prestressing steel, post- tension grouted strand, 100′ span, 100 kip	1,700	lb	$2.45	$1.39	$0.07	$3.91	$4.90
03 2305.50 0350	300 kip	3,200	lb	$2.09	$0.74	$0.04	$2.87	$3.49
Interpolating between the two lines:								
03230– 600–0325	200 kip	2,450	lb	$2.27	$1.07	$0.06	$3.40	$4.20

Area of concrete = 6 slabs × 120′ × 80′ = 57,600 SF
Weight of post-tension (PT) cable = 57,600 lb × 1 lb/SF = 57,600 lb

Bare cost = 57,600 lb × $3.40

= $195,840
≈ $196,000

Cost incl. O&P = 57,600 lb × $4.20 = $241,920 ≈ $242,000

Chapter 6 Exercises—Set A

For solutions to the exercises in Set A, see the link to the student companion website at www.wiley.com/go/constructionestim5e.

1. A 100-foot-long retaining wall with a cross section as shown in Figure 6.5 is to be constructed in Miami, Florida. Concrete will be pumped and will have a minimum compressive strength of 4,000 psi. Forms will be used four times. Estimate the total bare cost and the cost per linear foot for:

   A. Formwork

   B. Rebar

   C. Concrete

   D. All work

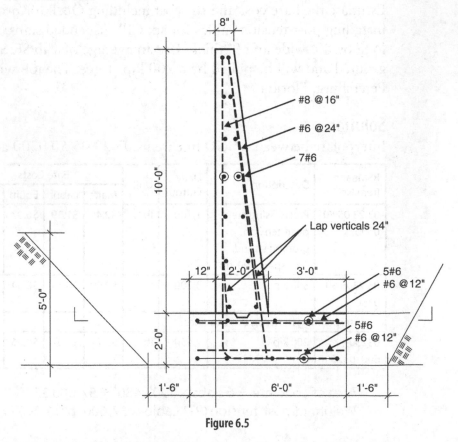

Figure 6.5

2. Estimate the total cost including O&P for the wall in Exercise 1, using RSMeans total including O&P.

3. If the wall in Exercise 1 will be built three years later, estimate the total bare cost based on an anticipated annual inflation of 5 percent.

4. Repeat Exercise 1, using these local labor wages:

Laborer	$28.00/hour
Labor foreman	$34.00/hour
Carpenter	$36.50/hour
Carpenter foreman	$40.00/hour
Rodman	$39.00/hour
Cement finisher	$36.00/hour
Equipment operator	$39.00/hour

5. Repeat Exercise 1 if the productivity drops by 20 percent, due to adverse weather conditions.

6. Repeat Exercise 1 if the pump will be rented on a monthly basis and used 20 days per month.

7. Repeat Exercises 1 and 2 using the shortcut (mini-assembly) method. (You may have to interpolate between two items.)

8. Repeat Exercise 1 using the forms five times.

9. Repeat Exercise 1 if workers will work 10 hours per day and 6 days per week. Overtime hours are compensated at 1.5 times the regular pay. Assume that this activity is part of an overall operation that will last two weeks. Make sure you reflect the effect of overtime on both productivity and hourly pay.

10. Estimate the total bare cost for constructing 40 columns and spread footings in New York City. Each column is 24″ × 24″ (square) and 10′ in clear height (to the bottom of the beam above). Footings are 4′ × 4′ in area and 2′ in depth. Forms will be used four times. Concrete will be pumped and will have a minimum strength of 3,500 psi. Reinforcement is as shown in Figure 6.6.

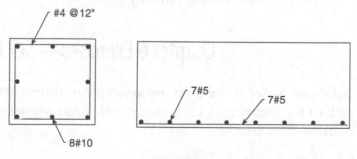

Figure 6.6

11. Estimate the total cost for the columns and footings, including O&P, in Exercise 11, using RSMeans total incl. O&P.

12. Repeat Exercise 10, using the local labor wages that are 50 percent higher than those mentioned in Exercise 4.

13. Repeat Exercise 10 without using the given reinforcement. Instead, use "average reinforcement" as shown in R033105–10.

14. Repeat Exercise 10 using the shortcut (mini-assembly) method. Consider the reinforcement average.

15. Repeat Exercise 10 using a crane and bucket for placing the concrete for columns.

16. Repeat Exercise 10 if the concrete pump will be rented on a daily basis.

17. Repeat Exercise 10 using the column forms five times.

18. Repeat Exercise 10 if crews are going to work 12 hours per day and 6 days per week. Overtime hours will be compensated at 1.5 times the normal pay. Assume that overtime will last two weeks only.

19. Repeat Exercise 10 if construction will take place after three years. Assume that inflation will be same as the average inflation between 2000 and 2010.

20. Estimate the bare cost to provide and install the following precast concrete members for a construction project in Portland, Oregon:

Member	Quantity
Rectangular beams, 30' span 18" × 44"	40 Ea
8"-thick slab plank, hollow	10,000 SF

21. Estimate the total cost including O&P for the precast members in Exercise 20.

22. What is the bare cost for the precast members in Exercise 20 if productivity is 15 percent better than that mentioned in the *BCCD* or RSMeans Online estimating software?

Chapter 6 Exercises—Set B

Solutions to Set B exercises are available to instructors only; see the link to the instructor's companion website at www.wiley.com/go/ constructionestim5e.

A three-story (four-level) concrete building to be used as a public garage in Frankfort, Kentucky, is the subject of the following exercises. The next plan represents a typical level. Other details follow (see Figure 6.7).

- The floor will be made of 8″-thick one-way suspended slabs, 12″ × 20″ beams (in the short direction), and 20″ × 30″ girders (in the long direction).

- Beam height is measured from the top of the slab.

- All columns will be square, 24″ × 24″, resting on spread (separate) footings; 4′ × 4′ × 2′.

- The floor-to-floor gross height will be 15′.

- The ground floor has a 6″-thick slab on grade with a 4″ × 4″ W4 × W4 wire mesh reinforcing and a 0.006″-thick polyethylene vapor barrier.

- On the top of the third floor (the roof), a 3′-high, 8″-thick CMU (concrete masonry unit) wall is to be built around the perimeter of the building.

- Use wood shores for slabs and beams.

- Formwork will be used four times.

- Steel grade is 60.

- Concrete minimum compressive strength is 3,500 psi and will be placed by a pump for all members.

- Assume reasonable concrete cover for rebar.

- State any other assumptions you have to make.

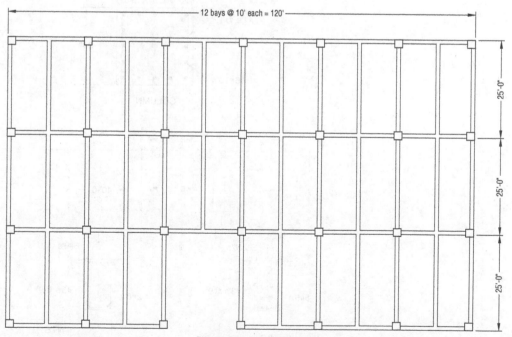

Figure 6.7 *(Continued)*

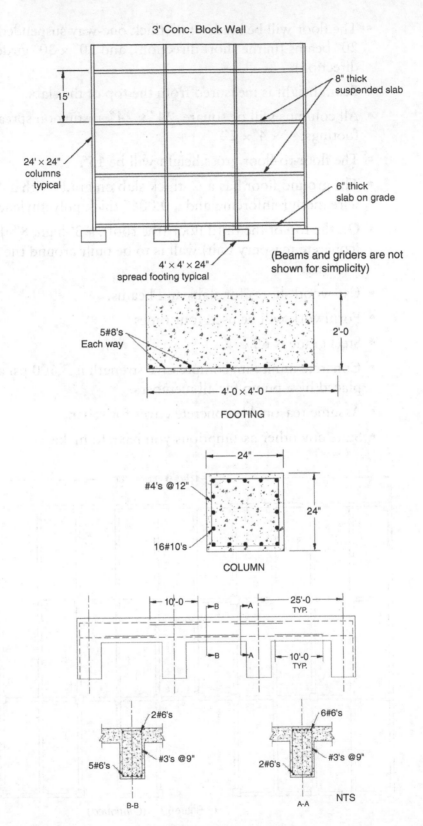

Figure 6.7 (*Continued*)

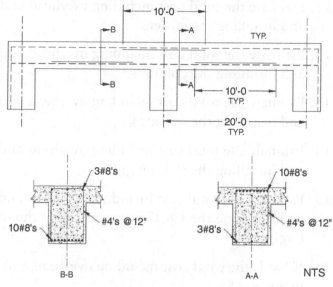

Figure 6.7 (*Continued*)

Perform the following:

1. Estimate the bare cost for providing and installing the suspended slabs.

2. Estimate the bare cost for providing and installing the (small) beams.

3. Estimate the bare cost for providing and installing the girders.

4. Estimate the bare cost for providing and installing the columns.

5. Estimate the bare cost for providing and installing the footings.

6. Estimate the bare cost for providing and installing the slab on grade.

7. Estimate the bare cost for providing and installing the CMU parapet wall on the top of the roof (third level).

8. What is the total bare cost for the project (items 1–7)?

For questions 9–16, use:

A. RSMeans total incl. O&P

B. RSMeans total bare cost plus labor add-ons and 10 percent of all expenses for overhead and profit. Check your answers by changing markups in RSMeans Online estimating, Advanced Report.

9. Estimate the total cost including overhead and profit for providing and installing the suspended slabs.

10. Estimate the total cost including overhead and profit for providing and installing the (small) beams.

11. Estimate the total cost including overhead and profit for providing and installing the girders.

12. Estimate the total cost including overhead and profit for providing and installing the columns.

13. Estimate the total cost including overhead and profit for providing and installing the footings.

14. Estimate the total cost including overhead and profit for providing and installing the slab on grade.

15. Estimate the total cost including overhead and profit for providing and installing the CMU parapet wall on the top of the roof (third level).

16. What is the total cost including overhead and profit for the project (items 9–15)?

17. Redo Exercise 1 assuming the concrete pump is to be rented on a monthly basis and there are 21 workdays per month.

18. Redo Exercise 1 using a crane and bucket for concrete placement.

19. Redo Exercise 1 using these labor prices:

Rodman	$35.00/hour
Cement finisher	$30.00/hour
Equipment operator	$31.00/hour
Laborer	$25.00/hour
Labor foreman	$30.00/hour
Carpenter	$31.00/hour
Carpenter foreman	$36.00/hour
Bricklayer	$34.00/hour
Bricklayer helper	$27.00/hour

20. Redo Exercise 1 if productivity is reduced by 20 percent.

21. Redo Exercise 8 if construction will take place during the year 2021. Assume 4 percent average annual inflation.

22. Redo Exercise 1 if work is being done 10 hours/day and 6 days/week. Assume a two-week duration and a 50 percent pay increase for overtime hours.

23. Redo Exercise 4 using the forms five times.

24. Redo Exercise 1 if reinforcement is not known. Use average reinforcement.

25. Redo Exercise 4 using steel-framed plywood forms.

Chapter 7

Masonry
Division 4

Types of Masonry

Masonry units include brick, concrete block, stone (both natural and artificial), and structural facing tile. They are available in many sizes, grades, colors, and textures. Prices differ significantly between one type or color and another.

Brick and block are used for structural (load-bearing) and other (aesthetic, insulation) non-load-bearing purposes. They are bonded by mortar, which is usually made by mixing Portland cement, lime, and sand with water. Metal ties, anchors, inserts, and horizontal reinforcement (truss or ladder type) are frequently added for more strength and stability. Concrete block walls are usually strengthened by grouting some cells and adding reinforcement bars in these cells (Figure 7.1).

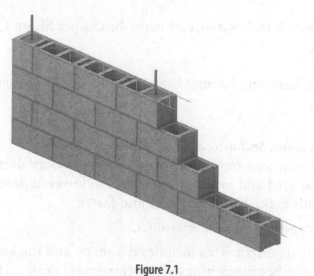

Figure 7.1

Estimating and Waste Allowances

When estimating the cost of structures built with masonry units, consult the project's plans and specifications to calculate materials and labor. Identify the types of masonry units, and calculate the quantity, including an allowance for waste (anticipating potential breakage). Masonry costs in RSMeans Online estimating and *Building Construction Cost Data* (*BCCD*) include a 3 percent waste allowance for masonry and a 25 percent waste allowance for mortar. (Note that waste applies to materials only.)

The number of masonry units needed is equal to the net area of the wall in square feet (SF) multiplied by the number of masonry units per SF, plus the waste allowance. The number of masonry units per SF is calculated from the brick size or can be obtained easily from Reference Number R042110–50 or from technical information supplied by manufacturers.

Nominal Brick Sizes

The nominal size of a masonry unit includes the mortar joint in both directions.

The normal mortar joint for brick is 3/8".

For a 4-inch-thick common brick, with nominal dimensions of $2\frac{2}{3} \times 8"$, the true dimensions are approximately $2.292" \times 7.625"$.

With the mortar joint, it will be $2\frac{2}{3}" \times 8"$, giving a face area of 21.333 square inches or 0.148 SF.

The number of bricks per SF = 1/(0.148) = 6.75 bricks per SF.

If the wall thickness is 8 inches, twice as many bricks per SF, or 13.5 bricks per SF, are required.

Productivity Factors

The productivity of installing (laying) bricks is influenced by several factors:

- *The type, size, color, and weight of the unit, and the kind of mortar and ties or other accessories used.* For instance, masonry units that are dark tend to be hard and brittle, which leads to lowered productivity; lightweight units make handling easier and faster.

- *The skill and experience of the mason(s).*

- *The complexity of design.* This includes the shape and thickness of the walls, the number of openings, the architectural design, the bond, special joint requirements, and the number of brick types involved.

- *Height and location of the work.* If the mason is working on scaffolding 15 feet above ground for example, productivity will be lower than working at ground level.

- *Quality of work and type of joint.* If the design allows for a small margin of tolerance and the owner has a strict quality assurance procedure, productivity will decrease. Exposed masonry requires well-manicured joints and high-quality finish and hence has lower productivity compared to masonry covered with materials such as stucco.

- *Other factors.* Other factors, such as weather conditions, job site congestion, and management–labor relationship, all affect productivity. (See "Productivity and Activity Duration" in Chapter 1.)

A reminder of two important points explained in Chapter 5:

1. Productivity fluctuations impact the unit price of crews: labor and equipment. It has no impact on materials unit cost.

2. Productivity fluctuations have a reverse proportional relationship with labor and equipment unit price. So, if productivity decreases by 20 percent, the unit price is divided by $(1 - 0.2)$; if it increases by 20 percent, the unit price is divided by $(1 + 0.2)$.

Quantity Takeoff

There are several methods for the quantity takeoff of masonry walls. Costs in RSMeans Online estimating and *BCCD* are given both per SF of the wall area and per M (1,000) bricks. The area of the wall is determined by the next equation:

$$\text{Area of wall} = [OSP - 4 \times Thk/12] \times Ht - (\text{Openings} > 10\,SF)$$

where:

$$OSP = \text{Outside perimeter of the building in feet}$$
$$Thk = \text{Thickness of the wall in inches}$$
$$Ht = \text{Height of the wall (building) in feet}$$
$$OSP - (Thk/12) = \text{Center-to-center perimeter of the exterior wall}$$

Deducting Openings

Frequently, openings less than 10 SF are not deducted. Some contractors do not deduct even larger openings, in exchange for not including the labor and materials needed to frame these openings. The cost of an opening, to a mason, includes the lintel and its installation as well as framing the opening. Overall, the cost of an average door or window opening is close to the cost of the additional block and its installation if that opening did not exist. (Many contractors will deduct only large openings that extend vertically all the way to the slab above, e.g., openings that don't need framing and have no lintel.) Typically, deductions are made for openings over 10 SF, and the cost of lintels and other items related to the opening is added.

When masonry walls (brick or concrete masonry units [CMU]) are long, control joints may be required. Any control joints must be added to the cost of the wall.

Example 1

Estimate the bare cost and total cost including overhead and profit for erecting a wall, 8′ high, 120′ long, in Milwaukee, Wisconsin. The wall will have 6″ concrete blocks (CMU) and 4″ face brick.

Solution

Area of wall = 120′ × 8′ = 960 SF. Use the following RSMeans line item:

RSMeans Item No.	Description	Unit	Bare Costs				Total Incl. O&P
			Mat.	Labor	Equip.	Total	
04 27 10.20 0400	Cavity wall, 4″ brick, 6″CMU	SF	6.60	14.47	–	21.07	29.50

Total bare cost = 960 SF × $21.07 = $20,227 ≈ $20,200

Total cost incl. O&P = 960 SF × $29.50 = $28,320 ≈ $28,300

Example 2

Estimate the bare cost of providing and installing common brick masonry for an 8″-thick outside wall of a rectangular building with outside dimensions of 40′ × 32′ and 11′ high to be built in New York City. Make proper deductions for the following openings:

Description	Dimensions	Quantity
Door	6′ – 8″ × 3′ – 0″	2
Door	6′ – 8″ × 2′ – 8″	1
Door	6′ – 8″ × 2′ – 6″	2
Window	4′ – 0″ × 4′ – 0″	4
Window	2′ – 0″ × 3′ – 4″	2
Window	4′ – 0″ × 5′ – 0″	3

Also, calculate the total number of bricks needed.

Solution

When we calculate the area of each of the openings in the preceding table, we find that the 2′ – 0″ × 3′ – 4″ windows are less than 10 SF each. Thus, they do not count in the deductions.

Description	Dimensions	Area, each	Quantity	Total Area SF
Door	$6' - 8'' \times 3' - 0''$	20	2	40
Door	$6' - 8'' \times 2' - 8''$	17.78	1	17.78
Door	$6' - 8'' \times 2' - 6''$	16.67	2	33.33
Window	$4' - 0'' \times 4' - 0''$	16	4	64.00
Window	$2' - 0'' \times 3' - 4''$	6.67	2	0*
Window	$4' - 0'' \times 5' - 0''$	20	3	60
			Total	215.11

*Note that the two small windows make more than 10 SF together, but we look at the individual opening size, not the total. Any opening smaller than 10 SF is not deducted.

To calculate area of wall:

$$\text{Outside perimeter} = 2 \times (40 + 32) = 144 \text{ LF}$$
$$\text{Area of wall} = (144 \text{ LF} - 4 \times 8/12) \times 11 \text{LF} = 1{,}555 \text{ SF}$$
$$\text{Deductions for openings} = 215 \text{ SF}$$
$$\text{Net wall area} = 1{,}555 \text{ SF} - 215 \text{ SF} = 1340 \text{ SF}$$

This calculation can be done per square foot of wall or per 1,000 bricks. For the square foot method, use the next RSMeans line number:

RSMeans Item No.	Description	Unit	Bare Costs				Total Incl. O&P
			Mat.	Labor	Equip.	Total	
04 27 10.30 0900	Wall, common brick, 8" thick	SF	8.59	24.00	–	32.59	46.09

$$\text{Total bare cost} = 1{,}340 \text{ SF} \times \$32.59 = \$43{,}671 \approx \$43{,}700$$

For the number of bricks required, use R042110–50.
Use standard brick, ⅜-inch joint.

$$\text{Number of bricks needed} = 1{,}340 \text{ SF} \times 13.5 \text{ bricks}/\text{SF}$$
$$= 18{,}090 \text{ bricks}$$

The contractor must add a waste allowance, say 3 percent. So, the order quantity = $18{,}090 \times 1.03 = 18{,}633$ bricks.

Bricks are not sold individually but rather supplied in cubes of 1,000 bricks. The contractor will probably order 19,000 or 20,000 bricks.

Using line 04 27 10.30 0204, the cost is $638.18 for materials and $1,795.36 for labor, or $2,433.54 for total bare cost (Figure 7.2).

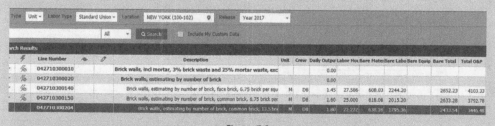

Figure 7.2

$$\text{Total bare cost} = 18.09 \, (\text{M bricks}) \times \$2,433.54 = \$44,023 \approx \$44,000$$

The difference between the two approaches, $461 (about 1 percent), is a result of rounding the unit cost. Calculating the cost per 1,000 bricks is slightly more accurate. Price in both items includes waste allowance for mortar and brick.

Example 3

Estimate the total cost including overhead and profit for the wall in Example 2.

Solution

Using item 04 27 10.30 0204:
The total cost including O&P is $2,125/M bricks (US national average) or $3,446.48 in New York City.

$$\text{Total cost} = 18.09 \, (\text{M bricks}) \times \$3,446.48 = \$62,347 \approx \$62,300$$

Example 4

Estimate the cost of furnishing and installing 10-inch regular exterior reinforced concrete blocks for a facility with a perimeter as shown in Figure 7.3. The wall is 12 feet high and has no openings except the entrance.[1] It will be built in Portland, Oregon. Cells are grouted with #5 bar in each cell.[2]

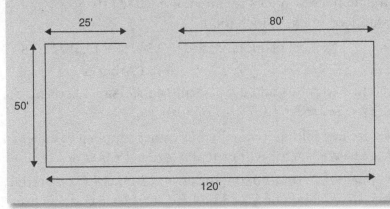

---

1. The building code would probably require several exit doors in the building, in addition to the main entrance, but we ignored them in this example for simplicity.
2. In most cases, some cells but not all are grouted and reinforced. For example, the structural designer can specify "grouting every 4th cell." In this case, the cost estimator has to pay attention to additional cells grouted at corners, door frames, and other locations.

Solution

Wall perimeter $= 25' + 80' + (2 \times 50') + 120' = 325\,\mathrm{LF}$

Wall area $= 325' \times 12' = 3,900\,\mathrm{SF}$

Number of blocks $= 3,900\,\mathrm{SF} \times 1.125$

$= 4,388$ blocks (*order quantity will add a waste factor* $\approx 3\%$)

Obviously, scaffolding will be needed for this height.
Use the following RSMeans items:

RS Means Item No.	Description	Unit	Bare Costs				Total Incl. O&P
			Mat.	Labor	Equip.	Total	
04 22 10.24 0250	Conc. block, ext, normal weight, 10″×8″×16″, includes mortar and horizontal joint reinforcing every other course, excludes scaffolding, grout, and vertical reinforcing	SF	4.52	5.87	–	10.39	13.97
01 54 23.70 0906	Scaffolding, steel tubular, regular, rent/month only for complete system for face of walls, 6′ – 4″ × 5′ frames, excl. planks	CSF	35.00	–	–	35.00	38.50
01 54 23.70 0090	Scaffolding, steel tubular, regular, labor only to erect and dismantle, building exterior, wall face, 6′ – 4″ × 5′ frames, 1 to 5 stories	CSF	–	146.37	–	146.37	223.51
01 54 23.70 2850	Scaffolding, steel tubular, regular, accessory, plank, rent/mo., 2′ × 10″ × 16′ long	EA	10	–	–	10	11
04 05 16.30 0300	Grout, CMU cores, 10″ thick, 0.340 CF/SF, pumped	SF	1.65	2.06	0.20	3.91	5.17
04 05 19.26 0060	Masonry reinf. bars, #5 and #6 bars, placed vertically	LB	0.49	0.57	–	1.06	1.40

For the scaffolding, the cost item chosen represents the rent cost per month. The productivity of the block item just described is 290 SF/day, so the expected duration of our job = 3,900/290 = 13.5 days. Actually, we should expect the duration to take a little more than that because of the height of the wall and the grouting and reinforcing of each cell. So, assume one month rent for the scaffolding.

For the planks, assume five planks across the scaffolding. This gives a width of almost 4 feet. (The actual width is 9.25 inches, not 10 inches. This point is discussed in Chapter 9.) Assume two levels of scaffolding, each at 5 feet high. Taking in consideration the length of the plank, 16 feet, and the dimensions of the wall, we need 23 lengths × 5 planks across the scaffolding × 2 levels = 230 planks.

For the rebar, the length of the perimeter is 325 feet. This means 244 blocks in one level, since the length of the block is 16 inches including the mortar joint. Each block has two cells, so there are 488 cells; each has a 12-foot-long steel bar.

$$\text{Total rebar} = 488 \times 12' \times 1.043\,\text{LB/LF} = 6{,}108\,\text{LB}$$

Total Bare Cost:

Block = 3,900 SF × $10.39	= $40,521
Scaffolding = 39 CSF × ($35.00 + $146.37)	= $ 7,073
Planks = 230 planks × $10	= $ 2,300
Grout = 3,900 SF × $3.91	= $15,249
Rebar = 6,108 × $1.06	= $ 6,475
Total bare cost	= $ 71,618

This is the equivalent of $18.36/SF, or $16.32/block.
Total cost including O&P:

Block = 3,900 SF × $13.97	= $54,483
Scaffolding = 39 CSF × ($38.50 + $223.51)	= $10,218
Planks = 230 planks × $11	= $ 2,530
Grout = 3,900 SF × $5.17	= $20,163
Rebar = 6,108 × $1.40	= $ 8,551
Total cost including overhead and profit	= $95,945
	≈ $96,000

This is the equivalent of $24.60/SF or $21.87/block.
Using the RSMeans Online Estimating with location adjustment is a huge advantage to the user over the manual adjustment method when using the *BCCD* print book with national average numbers and then adjusting with CCI codes. The user had to be very careful in using the correct CCI code, especially in "inter-division" items, such as wood items in the masonry or concrete divisions or metal items in the wood or masonry divisions.

Example 5

Estimate the cost of furnishing and installing decorative concrete blocks: 6-inch-thick, split-rib profile, 1-inch-deep ribs, 8 ribs per block. They will be installed on a four-story building with a total wall area of 12,400 SF in Fargo, North Dakota.

Solution

Use the next RSMeans items:

RSMeans Item No.	Description	Unit	Bare Costs				Total Incl. O&P
			Mat.	Labor	Equip.	Total	
04 22 10.23 5150	Conc blk, decor, split rib, 6" thick	SF	4.95	4.28	–	9.23	11.99
04 22 10.23 8550	High-rise constr., per story	MSF	–	20.41	–	20.41	31.40

To calculate the additional labor cost for high-rise construction:

$$\$20.41 \times 4 \text{ stories} = \$81.64/\text{MSF (thousand square feet)}$$

or

$$= \$81.64/\text{MSF} \times 12.4\,\text{MSF} = \$1,012$$

$$\text{Total bare cost} = 12,400\,\text{SF} \times \$9.23$$

$$= \$114,452$$

$$\text{Add for high rise} = \$1,012$$

$$\text{Total bare costs} = \$115,464 \approx \$115,500$$

Note that the extra $1,012 includes only the impact of the high rise and does not include cost of scaffolding. To add the cost of scaffolding, refer to Example 4 earlier.

Chapter 7 Exercises—Set A

For solutions to the exercises in Set A, see the link to the student companion website at www.wiley.com/go/constructionestim5e.

1. Estimate the bare cost for erecting a wall 14 feet high, 250 feet long, in Mobile, Alabama. The wall will be built of 8-inch-thick reinforced interlocking concrete blocks. Do not include scaffolding.

2. What is the order quantity for the job in Exercise 1?

3. Estimate the total cost including overhead and profit for the wall in Exercise 1 using RSMeans total cost including O&P.

4. Estimate the bare cost for the wall in Exercise 1 if the wall is to be 12 inches thick.

5. Estimate the bare cost for the wall in Exercise 1 if productivity is estimated at 70 percent of the normal level (i.e., 30 percent drop).

6. Estimate the bare cost for the wall in Exercise 1 if the local labor wages in Mobile are $33/hour for bricklayers and equipment operators and $27/hour for bricklayers' helpers.

7. Estimate the bare cost for the wall in Exercise 1 if workers will work 10 hours/day and 6 days/week for three weeks. Overtime hours are compensated at 1.5 times the regular pay.

8. Estimate the bare cost of providing and installing common brick masonry for an 8-inch-thick outside wall of a building with outside dimensions as shown in Figure 7.4. The wall height is 11 feet. The house will be built in Albuquerque, New Mexico. Make proper deductions for the openings shown in the following table.

Description	Dimensions	Quantity
Door	$6' - 8'' \times 3' - 0''$	3
Door	$6' - 8'' \times 2' - 8''$	2
Door	$2 \times [6' - 8'' \times 2' - 6'']$	1
Window	$4' - 0'' \times 4' - 0''$	4
Window	$2' - 0'' \times 3' - 4''$	2
Window	$4' - 0'' \times 5' - 0''$	3
Window	$2 \times [4' - 6'' \times 3' - 6'']$	1

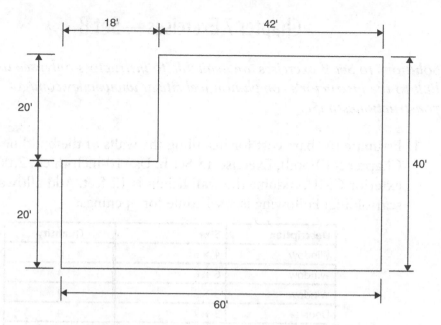

9. Calculate the order quantity of bricks for Exercise 8.

10. Estimate the total cost including overhead and profit for the building in Exercise 8 using RSMeans total cost including O&P.

11. Estimate the total cost including overhead and profit for the building in Exercise 8 using RSMeans total bare cost plus labor add-ons and 10 percent of all costs for O&P.

12. Assume the job in Exercise 8 is to be done during a hot and humid summer, which tends to cut down the productivity by 25 percent. Estimate the adjusted bare cost and total cost including overhead and profit using RSMeans total cost including O&P.

13. Redo Exercise 8, assuming the local wages in Albuquerque are $34/hour for bricklayers or equipment operators and $28/hour for bricklayers' helpers.

Chapter 7 Exercises—Set B

Solutions to Set B exercises are available to instructors only; see the link to the instructor's companion website at www.wiley.com/go/constructionestim5e.

1. Estimate the bare cost for installing the walls of the building in Chapter 9 (Wood), Exercise 15 Set B. Use 10-inch-thick, 2,000-psi exterior CMU. Assume the wall height is 12 feet. Add allowance for scaffolding. Following is a schedule for openings:

Description	Size	Quantity
Window	4′ × 6′	12
Window	6′ × 6′	4
Window	3′ × 3′	4
Door	3′ × 7′	4
Door	2′ × [3′-6″ × 8′]	2

2. What is the CMU order quantity for the job in Exercise 1?

3. Estimate the total cost, including overhead and profit, for the building wall in Exercise 1 using RSMeans total cost including O&P.

4. What is the estimated duration for the job in Exercise 1?

5. Estimate the bare cost for the wall in Exercise 1 if the wall is to be 12 inches thick.

6. Estimate the bare cost for the wall in Exercise 1 if productivity is estimated at 70 percent of the normal level.

7. Estimate the bare cost for the wall in Exercise 1 if the local labor wages in Las Vegas, Nevada, are $40/hour for bricklayers and equipment operators, $32/hour for bricklayers' helpers, and $38/hour for carpenters.

8. Estimate the bare cost for the wall in Exercise 1 if workers will work 10 hours/day and 6 days/week for three weeks. Overtime hours are compensated at 1.5 times the regular pay.

9. Estimate the bare cost of providing and installing 8-inch regular concrete block and then covering it with red brick veneer, rowlock course, for the outside perimeter for a house. The total length of the perimeter is 224 LF and the wall height is 10 feet. The house will be

built in Albuquerque, New Mexico. Make proper deductions for the openings shown in the following table.

Description	Dimensions	Quantity
Door	$6' - 8'' \times 3' - 0''$	2
Door	$2 \times [6' - 8'' \times 2' - 6'']$	1
Window	$4' - 0'' \times 4' - 0''$	4
Window	$2' - 0'' \times 3' - 4''$	3
Window	$4' - 0'' \times 6' - 0''$	3
Window	$2 \times [4' - 6'' \times 3' - 6'']$	1

10. Calculate the order quantity of blocks and bricks for Exercise 9. Add suitable waste allowance.

11. Estimate the total cost including overhead and profit for the job in Exercise 9 using RSMeans total cost including O&P.

12. Estimate the total cost including overhead and profit for the job in Exercise 9 using RSMeans total bare cost plus labor add-ons and 10 percent of all costs for O&P.

13. Assume the job in Exercise 9 is to be done during a hot and humid summer, which tends to cut down the productivity by 25 percent. Estimate the adjusted bare cost and total cost including overhead and profit using RSMeans total cost including O&P.

14. Redo Exercise 9, assuming the local wages in Albuquerque are $34/hour for bricklayers or equipment operators and $28/hour for bricklayers' helpers.

Chapter 8

Metals
Division 5

Metals are used extensively in construction due to their high strength-to-weight ratio. Steel is commonly used for the structural system of buildings. Metals also appear throughout buildings as connectors, metal decking, cold-rolled metal framing, and miscellaneous metal fabrications, including railings, grilles, and ladders.

Estimating Structural Steel

Structural steel is normally purchased already fabricated to project specifications. Erection may be done by the general contractor (GC) or a subcontractor. After the GC submits the design (drawings and specifications) to the fabricator, the fabricator prepares a takeoff list of all structural members, details, and miscellaneous items. Costs added to the base price of fabrication may include extra charges for quantity, size, cut-length, milling, galvanizing, and other items that will be detailed later. The fabricator also adds all labor costs associated with preparing and delivering the steel, such as drafting, fabricating, shop welding, shop painting, transportation, and shop overhead and profit.

Installation costs are calculated separately from fabrication costs. Installation costs include unloading, sorting, storing, erection, field-bolting or welding, field-painting, equipment costs, and job and general overhead and profit. There should be a clear distinction between fabrication labor and erection labor charges and between shop overhead and profit and erector's overhead and profit.

The detailed cost estimate for structural steel is often prepared by a subcontractor and supplied to the contractor. The structural steel cost data in RSMeans Online estimating and *Building Construction Cost Data (BCCD)* can be useful to the GC wishing to verify a subcontractor's price by doing his or her own estimate. Costs include fabrication of members, delivery to the site, and erection. In many cases, only main structural members are priced, and an allowance is made for base plates, connections, column splices, and other details. This allowance may vary from 10 to 20 percent of the gross tonnage, depending on the type of project and the estimator's judgment and experience.

Types of Structural Steel Sections

Structural steel consists of beams, columns, plates, joists, and decking. Common sections and their designations are detailed and defined in Reference Number R051223–35. The types of steel alloys are defined in R051223–25 and R051223–30.

Structural Steel Extra Costs

The costs provided in RSMeans Online estimating and the *BCCD* are for A36/A992 structural projects ordered in quantities of 100 tons or more. Costs are for common shapes—fabricated, delivered to the site, and erected. The bare material price per ton of normal A36/A992 steel, for projects of 100 tons or more, fabricated and delivered to a job site, is shown on line 05 12 23.77 3880. The breakdown of costs that make up the cost of this line are shown on lines 05 12 23.77 3810 through 3870. Many factors influence the total price of structural steel, which may be calculated as extras to the base price:

- *High strength*. High-strength steel may cost more per ton than the basic A36/A992, but overall savings may be achieved through reduced weights. Designers should consult with fabricators and cost estimators for the optimum design scheme.

- *Quantity*. The price increases if the quantity of *any certain section* is less than 5 tons (10,000 lb) or if the total weight for the job is less than 100 tons.

- *Size*. Commonly used W sections do not bear any extra cost. All other section types do. Section 05 12 23.40 lists the cost of several lightweight framing sections.

- *Cambering*. Cambering is giving the beam an inverted (upward) deflection so that when loads are applied, the total deflection is very small. Fabricators charge extra for cambering.

- *Galvanizing*. This term refers to the process of coating steel with a thin layer of zinc for protection against corrosion. It is required or recommended for sections exposed or vulnerable to corrosion. Section 05 05 13.50 lists the cost of galvanizing structural steel in shop, which ranges from $450 to $540/ton, depending on quantity. For paints and protective coatings including cold galvanizing in the field, the cost is detailed under Section 09 97 13.23.

- *Cut lengths*. There is an additional charge for short lengths from the mill (less than 25 feet) and for long lengths (greater than 90 feet). Section 05 05 21.10 shows the cost of cutting steel.

- *Drilling*. Drilling for bolts or rivets, or other uses costs extra charge. Refer to Section 05 05 21.15.

- *Other extras*. There are additional charges for government specifications, milling ends, handling and loading, special testing, special straightening, splitting beams to produce tees, and chemical testing.

When using Division 5 in RSMeans Online estimating and *BCCD*, it's important to know what extras are already included. For example, items 05 12 23.75 (0100 through 8100) are based on projects of 100 tons or more. If the total weight for the job is less than 100 tons, an extra percentage must be added to the cost of materials, as per items 05 12 23.75 8490 through 8499. The beams in Section 05 12 23.75 do not bear any size extra. For extra costs, refer to Sections 05 12 23.77 3900 through 4100 for high-strength extras and to Sections 05 12 23.77 4200 through 4230 for mill spec extras. The material cost includes an average cut length extra. Lintels, items 05 12 23.45 0010 through 0300, are already adjusted for quantities.

Connections for structural steel members are done using bolts or welding. Welding is considered better, in general, than bolting from a structural point of view. This is because welding does not require the drilling of holes, which reduce the section area, thus weakening it, as done in the case of bolting. Welding also cost more than bolting. Section 05 05 21.90 addresses the cost of welding. There are items in the *BCCD* that mentions simple connections or moment / composite connections. Simple connections are bolted with either A325 or A490 high strength bolts. Moment/composite connections are first bolted simply to hold the members together in proper position, and then they are welded in accordance with the stamped Structural Drawings & Notes.

Per-Ton Estimates: Sections 05 12 23.77 0200 through 3800 give estimates for structural steel projects per ton of weight. The numbers are for an average 100-ton job. R050516–30 can be used as a general guideline for the *surface area per ton* ratio. Line 05 12 23.77 0700 shows the cost to construct a one- or two-story office building. Items 05 12 23.77 3810 through 5399 give adjustments for various factors such as mill extras plus delivery to warehouse, delivery from warehouse to fabrication shop, shop extra for drawings and detailing, shop extra for fabrication and handling, mill spec extras, mill size extra, jobs under 100 tons, and other. It is important for the user to read the item description carefully and then search other items for any extra cost.

Pre-engineered Steel Buildings: Pre-engineered steel buildings are used frequently for warehouses and industrial buildings due to their wide spans (up to 200 feet) and economy. These buildings are manufactured by specialty manufacturers and erected by licensed contractors. Pre-engineered steel buildings and other projects (e.g., silos, broadcast towers, or greenhouses) are priced in RSMeans Division 13, Special Construction, Section 13 34 19.50, based on span length, eave height, and other criteria.

Estimating Steel Members

For W members[1] (beams and columns), the first number denotes the nominal depth of the member in inches. Actual depths vary, so a certain nominal depth includes sections with actual depth bigger than their nominal depth by a fraction of an inch to a few inches and sections with actual depth smaller than their nominal depth by a fraction of an inch. The second number is the weight per lineal foot in pounds. For example, the section W 18×50 is about 18 inches (17.99 inches) deep and weighs approximately 50 PLF (pounds per lineal foot). The sections W 18×119 and W 18×35 have depth of 18.97 and 17.7, respectively, and weigh approximately 119 and 35 PLF, respectively. The actual weight may vary slightly. The cost will be based on actual weight. The tables are also available in metric units.

The weight per lineal foot for steel sections W, M, S, HP, standard channels (C), and miscellaneous channels (MC) is represented by the second number in the section ID, as mentioned earlier. Other sections, such as angles, pipes, and others, can be found in steel manuals. For example, section L 4×4×½ weighs about 12.8 PLF. For base plates or any solid steel section, the weight can be found in a steel table or calculated by multiplying the volume by the unit weight, 490 PCF (pounds per cubic foot). WT sections (structural Ts) are cut from W shapes. For example, WT 10.5×46.5

---

1. The term *W-section* came originally from their description as "wide flange" steel sections. The terms *I-beams* and *H-columns* have been also used as the steel beams mimic the letter I while the steel columns mimic the letter H. They both are W sections, but those used as beams usually have depth much bigger than the flange width for better resisting the bending moment, while those used as columns have depth and flange width close in dimensions for better resisting the vertical compressive loads including buckling.

has an actual depth of 10.810 inches and an actual weight of 46.5 PLF. Comparing it with W 21×93, you'll find the latter WT section has exactly twice the depth and twice the weight per linear foot.

In RSMeans Online estimating and *BCCD*, costs given are for structural steel sections, including bolted connections, unless otherwise noted. Welded connections cost more. Refer to Sections 05 05 21.90, 05 05 23.85, 05 05 23.87, and 05 05 23.90 and to Reference Number R050521–20. High strength bolting also cost more than regular bolted connections. See section 05 05 21.90 for example.

Example 1

Using RSMeans Online estimating, estimate the total bare cost for the main structural steel members used for framing one story in the rectangular office building shown in Figure 8.1. The project will be in Phoenix, Arizona. Assume columns are 20 feet high.

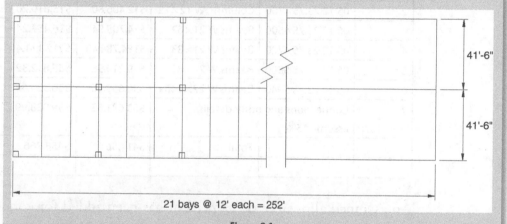

21 bays @ 12' each = 252'

Figure 8.1

Solution

Item No.	No. of Pieces	Description	Length Ea	Weight lb/LF	Total Length (LF)	Total Weight (lb)
1	18	Columns, W 12×50	20.0	50	360	18,000
2	6	Columns, W 12×72	20.0	72	120	8,640
					Total weight of columns	26,640
3	4	Beams W 21×62	41.5	62	166	10,292
4	40	Beams W 21×83	41.5	83	1,660	137,780
5	14	Beams W 27×84	36.0	84	504	42,336
6	7	Beams W 27×146	36.0	146	252	36,792
					Total weight of beams	227,200
					Total weight of fabricated steel	253,840 lb
						126.92 tons

Item No.	CSI Number	Description*	Crew	Daily Output	Unit	Unit Price				
						Bare Mat.	Bare Labor	Bare Equip.	Bare Total	Total Incl. O&P
1	05 12 23. 75 1560	Columns, W 12×50	E2	750	LF	$67.46	$2.97	$2.09	$72.52	$81.63
2	05 12 23.75 1700	Columns, W 12×72	E2	640	LF	$97.45	$3.49	$2.45	$103.39	$115.16
3	05 12 23.75 4500	Beams W 21×62	E5	1,036	LF	$83.86	$3.10	$1.63	$88.59	$99.14
4	05 12 23.75 4720	Beams W 21×83	E5	1,000	LF	$112.44	$3.21	$1.69	$117.34	$130.79
5	05 12 23.75 5800	Beams W 27×84	E5	1,190	LF	$113.38	$2.69	$1.42	$117.49	$130.58
6	05 12 23.75 5940	Beams W 27×146	E5	1,150	LF	$196.77	$2.79	$1.47	$201.03	$223.57

Item No.	CSI Number	Description	Bare Total	Total Incl. O&P
1	05 12 23.75 1560	Columns, W 12×50	$26,107.20	$29,386.80
2	05 12 23.75 1700	Columns, W 12×72	$12,406.80	$13,819.20
3	05 12 23.75 4500	Beams W 21×62	$14,705.94	$16,457.24
4	05 12 23.75 4720	Beams W 21×83	$194,784.40	$217,111.40
5	05 12 23.75 5800	Beams W 27×84	$59,214.96	$65,812.32
6	05 12 23.75 5940	Beams W 27×146	$50,659.56	$56,339.64
7	Connections and other details, assume 15%		$53,681.83	$59,838.99
		Total	$411,561	$458,766

An assumed allowance of 15 percent has been added for connections and other details. Estimators may assume different percentages based on the type of project.

When we do the same estimate using RSMeans Online estimating, all numbers are automatically adjusted for location, Phoenix, Arizona. We obtain the following results.

Note that members in Section 05 12 23.17 (columns) can be used with concrete, masonry, and even wood structures, in addition to steel structures. Section 05 12 23.75 (structural steel members) includes both beams and columns, and is mainly for use for structural steel buildings.

*All item descriptions contain: "Structural steel member, 100-ton project, one- to two-story building, A992 steel, shop fabricated, including shop primer, bolted connections."

Line Number	Long Description	Quantity	Unit of Measure	Extended Equipment	Extended Labor	Extended Material	Extended Total	Extended Total OP
051223751560	Structural steel beam or girder, 100-ton project, 1 to 2 story building, W12 x 50, A992 steel, shop fabricated, incl shop primer, bolted connections	360.00	L.F.	$752.40	$1,069.20	$24,285.60	$26,107.20	$29,386.80
051223751700	Structural steel beam or girder, 100-ton project, 1 to 2 story building, W12 x 72, A992 steel, shop fabricated, incl shop primer, bolted connections	120.00	L.F.	$294.00	$418.80	$11,694.00	$12,406.80	$13,819.20
051223754500	Structural steel beam or girder, 100-ton project, 1 to 2 story building, W21 x 62, A992 steel, shop fabricated, incl shop primer, bolted connections	166.00	L.F.	$270.58	$514.60	$13,920.76	$14,705.94	$16,457.24
051223754720	Structural steel beam or girder, 100-ton project, 1 to 2 story building, W21 x 83, A992 steel, shop fabricated, incl shop primer, bolted connections	1,660.00	L.F.	$2,805.40	$5,328.60	$186,650.40	$194,784.40	$217,111.40
051223755800	Structural steel beam or girder, 100-ton project, 1 to 2 story building, W27 x 84, A992 steel, shop fabricated, incl shop primer, bolted connections	504.00	L.F.	$715.68	$1,355.76	$57,143.52	$59,214.96	$65,812.32
051223755940	Structural steel beam or girder, 100-ton project, 1 to 2 story building, W27 x 146, A992 steel, shop fabricated, incl shop primer, bolted connections	252.00	L.F.	$370.44	$703.08	$49,586.04	$50,659.56	$56,339.64
	Structural steel allowance (Connections and other details on Extended Total OP), 15%						$53,681.83	$59,838.99
						Grand Total	$411,561	$458,766

Cost for members in this section does not include details (which is why we added 15 percent in our previous example).

Most structural steel contractors calculate the total weight for structural steel, then multiply by a *rounded average* price per ton. In our example, the cost per ton will be:

$$\text{Bare cost per ton} = \$411,561/146^2 \text{ tons} = \$2,820$$
$$\text{Total cost incl. O\&P per ton} = \$458,766/146 \text{ tons} = \$3,143$$

The numbers may also be presented as $19.68/SF (bare cost) or $21.93/SF (including O&P)

Example 2

Calculate the bare cost and total cost including O&P for sections W 21×83 in Example 1 using the *BCCD*.

Solution

Using line item 05 12 23.75 4720:

RSMeans Item No.	Description	Unit	Bare Costs				Total Incl. O&P
			Mat.	Labor	Equip.	Total	
05 12 23.75 4720	Struc. Steel, bolted, W 21×83	LF	$112.44	$3.21	$1.69	$117.34	$130.79

$$\text{Bare cost} = 1,660 \text{ LF} \times \$117.34 = \$194,784 \approx \$195,000$$
$$\text{Total cost incl. O\&P} = 1,660 \text{ LF} \times \$130.79 = \$217,111 \approx \$217,000$$

To obtain the total cost including O&P using the *BCCD* bare cost plus add-ons, use the next calculations:

Labor add-ons:

Workers' compensation insurance, struct. steel, Arizona:	24.4%×89%=21.72%
Average fixed overhead	18.3%
Overhead	14.00%
Profit	10.00%
Total	64.02%

---

2. This number includes 15 percent additional to allow for the connections and other details.

Total cost including O&P for W 21×83 beams:

Materials = 1,660 LF × $112.44	= $ 186,650
Materials profit = $191,018 × 10%	= $ 18,665
Labor = 1,660 LF × $3.21	= $ 5329
Labor add-ons and profit = $5,385 × 64.02%	= $ 3,412
Equipment = 1,660 LF × $1.69	= $ 2,805
Equipment profit = $2,554 × 10%	= $ 281
Total	= $ 217,142
	≈ $ 217,000

The two results are almost identical.

Example 3

If the steel in Example 1 was to be used for a one-story office building in Phoenix, Arizona, using the quantities calculated earlier, estimate the bare cost and the total cost including overhead and profit.

Solution

Use 05 12 23.77 0700:

$$\text{Total bare cost} = 146 \text{ tons} \times \$2{,}934.16 / \text{ton} = \$428{,}387$$
$$\text{Total cost including O\&P} = 146 \text{ tons} \times \$3{,}383.89 / \text{ton} = \$494{,}048$$

Example 4

If the building in Example 3 had four stories and field-welded connections were used, what would be the total bare cost and cost including O&P?

Solution

Use item 05 12 23.77 0800, offices, etc. 3–6 stories, to find the new base price for building height:

RSMeans Item No.	Description	Unit	Bare Costs				Total Incl. O&P
			Mat.	Labor	Equip.	Total	
05 12 23.77 0800	Offices, etc. 3–6 stories	LF	2,506.48	354.72	129.44	2,990.64	3,489.60

To determine the additional cost for welded connections, compare lines 05 12 23.77 0700 and 3400.

RSMeans Item No.	Description	Unit	Bare Costs				Total Incl. O&P
			Mat.	Labor	Equip.	Total	
05 12 23.77 0700	Offices, etc, steel, 1–2 stories	LF	2,459.63	310.38	164.15	2,934.16	3,383.89
05 12 23.77 3400	Welded const, 1–2 stories	LF	2,506.48	421,23	238.25	3,165.96	3,716.82
	Difference $/ton	LF	46.85	110.85	74.10	231.80	332.93

$$\text{Total bare cost} = \$2,990.64 + \$231.80 = \$3,222.44/\text{ton}$$
$$\text{Total cost incl. O\&P} = \$3,489.60 + \$332.93 = \$3,822.53/\text{ton}$$

For a 146-ton building:

$$\text{Total bare cost} = 146\text{T} \times \$3,222.44/\text{ton} = \$470,476$$
$$\text{Total cost incl. O\&P} = 146\text{T} \times \$3,822.53/\text{ton} = \$558,089$$

Example 5

Estimate the cost for the roof structure on a rectangular building, 230 feet long by 57 feet wide in Arlington, Virginia. The roof is made up of bar joists, 22K9 running in the short direction at 5 feet OC and a 3-inch, 20-gauge, open-type, galvanized steel deck. There is a 1:12 slope with 1-foot overhang in each direction. The roof will be supported by a load-bearing masonry wall (not included in the estimate).

Solution

Total building area	$= 230' \times 57' = 13,110 \text{ SF}$	$= 131.1 \text{ SQ}$
Roof factor	$= \sqrt{1 + S^2}$	$= 1.0035$
Total roof area	$= 232' \times 59' \times 1.0035$	$= 13,736 \text{ SF}$
Number of joists needed	$= (230 / 5') + 1$	$= 47 \text{ joists}$
Total length of needed joists	$= 47 \text{ joists} \times 57' \text{ each}$	$= 2,679 \text{ LF}$

Use the following RSMeans line items:

RSMeans Item No.	Description	Unit	Bare Costs				Total Incl. O&P
			Mat.	Labor	Equip.	Total	
05 21 19.10 0560	Open web joist, 22K9	LF	10.04	2.27	1.05	13.36	15.95
05 31 23.50 3360	Metal roof decking, steel, over 500 Sq, 3" D, 20 gauge	SF	2.61	0.48	0.03	3.12	3.73

Bare cost:

$$\text{Bar joists} = 2,679\,\text{LF} \times \$13.36 = \$35,791$$
$$\text{Steel deck} = 13,736\,\text{SF} \times \$3.12 = \$42,856$$
$$\text{Total bare cost} = \$78,647 \text{ or } \$5.73/\text{SF of the roof area}$$

$$\text{Total cost incl. O\&P} = (2,679\,\text{LF} \times \$15.95) + (13,736\,\text{SF} \times \$3.73)$$
$$= \$93,965 \text{ or } \$6.84/\text{SF of the roof area}$$

Example 6

Estimate the cost of the structural steel needed to erect a 10-story office building in downtown Minneapolis, Minnesota. Each story is about 12,500 SF in gross area.

Solution

There is very little information given here. Our estimate will be an approximate one. We can use line item 05 12 23.77 0900 (with adjustments). The price is given per ton of weight. Since we don't have the weight, we have to estimate it based on the building area. Assume 12 lb/SF. This assumption is made based on previous experience and can vary with different projects.[3]

$$
\begin{aligned}
\text{Total area} &= 10 \text{ stories} \times 12,500\,\text{SF}/\text{story} &&= 125,000\,\text{SF}\\
\text{Total estimated weight} &= 125,000\,\text{SF} \times 12\,\text{lb}/\text{SF} &&= 1,500,000\,\text{lb}\\
&= 1,500,000\,\text{lb}/2,000 &&= 750 \text{ tons}
\end{aligned}
$$

Using line item 05 12 23.77 0900, total bare cost = $3,285.35/ton and total cost including O&P = $3,934.37/ton.

$$
\begin{aligned}
\text{Bare cost} &= 750 \text{ tons} \times \$3,285.35/\text{ton}\\
&= \$2,464,013 \approx \$2,464,000\\
\text{Total cost incl. O\&P} &= 750 \text{ tons} \times \$3,934.37/\text{ton}\\
&= \$2,950,778 \approx \$2,951,000
\end{aligned}
$$

3. Review Note R051223-20 Steel Estimating Quantities.

Chapter 8 Exercises—Set A

For solutions to the exercises in Set A, see the link to the student companion website at www.wiley.com/go/constructionestim5e.

1. Estimate the bare cost for the structural steel members needed for a framed steel building in Atlanta, Georgia. All members are to be fabricated at a shop for bolted connections. Use regular strength (A992) steel. A list of the main members is given next. Add 15 percent of the total cost for details.

Quantity	Description	Length (Ea)	Total Length (LF)	Total Weight (lb)
12	Columns, W 8×48	20'-6"		
24	Columns, W 10×68	20'-6"		
		Total weight of columns: _____		
60	Beams W 10×15	19'-6"		
36	Beams W 14×34	25'-0"		
18	Beams W 18×55	26'-6"		
		Total weight of beams: _____		
		Total weight of fabricated steel (lb): _____		
		Total weight of fabricated steel (tons): _____		

2. Assume the structural steel members in Exercise 1 are used for an industrial building in Atlanta. Use the total quantity (tons) from Exercise 1 (plus 15 percent details) to estimate the total bare cost and cost including O&P of the building, using the *BCCD* Structural Steel Projects (Section 05 12 23.77).

3. Redo Exercise 2, assuming the building is three stories high and field-welded connections are used.

4. Redo Exercise 2 with the contractor owning the crane. It costs the contractor an average of $680/day in direct ownership expenses (operating expenses not included).

5. Redo Exercise 2 if the building is to be built during the year 2019. Assume an average annual inflation of 5 percent.

6. A warehouse in Miami, Florida, will be built of structural steel. Estimate the bare cost of the roof if trusses are used and the steel roof trusses weigh 6 tons. Add a galvanized metal deck, cellular units, 3 inches deep, 16–18 gauge. Roof size is 8,400 SF. Use average cost for trusses.

7. Redo Exercise 6 with an experienced crew working under favorable conditions, resulting in an expected increase in productivity of 15 percent.

8. Redo Exercise 6 with the crew working 11 hours per day, 6 days per week. Assume a 2-week duration and 1.5 times the regular pay for overtime hours.

9. Redo Exercise 6 using the following labor wages:

Structural steel foreman	$40/hour
Structural steel worker	$35/hour
Equipment operator	$36/hour
Welder	$35/hour
Equipment operator oiler	$31/hour

10. A heavy machine needs to be placed on the second floor of an old building in Omaha, Nebraska. In order to give the floor adequate support, we need to add 12 columns of extra-strong galvanized steel, 6 inches in diameter, filled with concrete. Each column is 10 feet high. The weight of the steel is 28.57 lb/LF. Estimate the total bare cost.

11. Estimate the total cost including O&P for the columns in Exercise 10.

12. Estimate the bare cost for a 150-ton, structural steel, one-story industrial building in Boston, Massachusetts, with masonry bearing-wall construction and field-welded connections. Assume a cost multiplier of 96 percent, because this project is larger than the average.

13. Estimate the total cost including O&P for the building in Exercise 12.

14. Redo Exercise 12 using A588 steel.

Chapter 8 Exercises—Set B

Solutions to Set B exercises are available to instructors only; see the link to the instructor's companion website at www.wiley.com/go/constructionestim5e.

1. Estimate the bare cost per ton of structural steel for the columns, beams, and details for a framed steel building in Rochester, New York. All members are to be fabricated at a shop for bolted

connections. Use regular strength (A992) steel. The list of members and details is given next.

Quantity	Description	Length (ea)	Total Length (LF)	Total Weight (lb)
10	Columns, W 8×31	15'-6"		
20	Columns, W 12×87	17'-0"		
20	Columns, W 12×120	17'-0"		
			Total weight of columns: _____	
180	Ls 6×6×½	1'-2" (19.6 lb/ft)		
150	Ls 4×4×3/8	1'-0" (9.8 lb/ft)		
30	Base plates, 1" thick	2'×2'	163 lb, each	
20	Base plates, 2" thick	2'×2'6"	408 lb, each	
			Total weight of columns details: _____	
22	Beams W 10×26	18'-0"		
28	Beams W 10×49	22'-0"		
36	Beams W 16×50	24'-3"		
24	Beams W 21×93	29'-6"		

Total weight of beams: _____

Total weight of fabricated steel, lb: _____

Total weight of fabricated steel, tons: _____

2. Estimate the total cost including overhead and profit for the structural steel in Exercise 1.

3. Assume the structural steel members in Exercise 1 are used for a five-story office building in Rochester, New York. Use the total quantity (tons) above to estimate the total bare cost and cost including O&P of the building, using RSMeans Structural Steel Projects (Section 05 12 23.77).

4. Redo Exercise 3 using RSMeans line 05 12 23.77 3400 for welded connections.

5. Redo Exercise 3 using these labor wages:

Structural steel foreman	$50.00/hour
Structural steel worker	$45.00/hour
Equipment operator	$40.00/hour
Equipment operator oiler	$36.00/hour
Welder	$46/hour

6. Redo Exercise 3 with the assumption that the contractor owns the crane and it costs $920/day in direct ownership cost.

7. Redo Exercise 3 if construction is to be done in 2019. Assume an average inflation equal to the one between 1995 and 2015.

8. Estimate the bare cost for providing open web joists, 22K9. They will be used to roof a rectangular building that measures 40 feet wide by 120 feet long. Joists will be spaced at 5 feet OC. The building is in Charlotte, North Carolina.

9. Estimate the total cost including O&P for the joists in Exercise 8.

10. Estimate the bare cost for a 90-ton structural steel, four-story office building in Phoenix, Arizona, with masonry bearing wall construction and field-welded connections. Use Division 5 (Structural Steel Projects) and line 05 12 23.77 3400.

11. Estimate the total cost including O&P for the building in Exercise 10.

12. Redo Exercise 10 assuming it is to be done between the end of 2020 and early 2021.

13. Redo Exercise 10 using A572 steel, grade 60.

Chapter 9

Wood and Plastics, Thermal and Moisture Protection
Divisions 6 and 7

Wood and Wood Products

Wood possesses many properties that make it suitable for a wide variety of uses in construction. It has a high degree of strength relative to its weight during compression, tension, and bending. It is resistant to impact and can be worked easily to desired shapes with simple tools. Prized for its decorative character, it also has an almost infinite variety of grains, figures, textures, colorings, and markings. Many varieties are valued as a renewable natural resource.

Woodwork is divided into these categories:

- Rough carpentry, including framing, sheathing, laminated wood products, and trusses
- Finish carpentry, including moldings, shelving, and paneling
- Architectural woodwork, including cabinets, drawers, and ornamental items

In addition to wood products, MasterFormat Division 6, Wood and Plastics, also includes pricing information on structural plastics and plastic fabrications, such as fiberglass gratings, castings, railings, and stair parts.

Wood products are also included in line items in other divisions, such as concrete formwork in Division 3, wood doors and windows in Division 8, wood flooring in Division 9, and many others.

Nominal versus Real Dimensions

It is important to note that the actual dimensions for a dressed member are *smaller* than the corresponding nominal dimensions for that member. Conversely, real dimensions for a rough member are slightly *larger* than those of a dressed member of the same nominal size.

Nominal dimension lumber of 6 inches or less has a dressed real dimension of ½ inch less than the nominal dimension. Any nominal dimension lumber of greater than 6 inches has a dressed real dimension of ¾ inch less than the nominal dimension. For example, a 2 × 4 is actually dressed to 1½ inches by 3½ inches and a 2 × 8 is actually dressed to 1½ inches by 7¼ inches. All timbers have a dressed size of ½ inch less than the nominal dimension. An 8 × 8 post is actually 7½ inches by 7½ inches.

Those renovating old and historic buildings may see members of sizes that are no longer available in the market today.

All quantities for the purpose of cost estimating (quantity takeoff, ordering, billing, etc.) must be calculated based on nominal dimensions.

Lumber Quantities

Lumber is measured and sold by the board foot. A board foot (BF) is the volume of a square board with a length and width equal to 1 foot each, and a thickness of 1 inch. Lumber members are usually identified by their cross-sectional dimensions (width and depth) in inches and their length in feet.

Thus, to calculate the volume of a lumber piece in board feet, multiply the three dimensions, inches × inches × feet, and divide by 12. A 16-foot-long 2 × 12 member is (2″ × 12″ × 16′) / 12 = 32 BF.

Because a board foot is a relatively small unit, it is common to estimate using the unit MBF (1,000 board feet).

Lumber Waste

Unlike other materials, lumber waste cannot be assumed as a simple fixed percentage or a rule of thumb. Waste in lumber consists of two elements:

1. *Design waste.* All pieces must be ordered in increments of two feet. If the design calls for a 16′-8″ long member, the order length must be 18 feet. In this case, the waste is 1′-4″, or 8 percent.[1]

2. *Regular waste.* Pieces that are broken, damaged, or mistakenly cut shorter than desired are considered waste. Defective pieces may

---

1. The author had an interesting situation while doing cost estimating to a large house. The joists were supposed to be 14′-2″ long. The author suggested that the partition be moved 2 inches so the joists would be exactly 14 feet. The owner did not object and was delighted to save a bundle by buying 14-foot joists instead of 16-foot long; saving in both material and cutting cost.

be considered waste, or they may be returned to the vendor for replacement.

As a rule of thumb, regular waste can be estimated at up to 3 percent. However, design waste has to be calculated on a case-by-case basis. The total waste for lumber is usually between 3 and 15 percent.

Plywood waste depends on building shape and dimensions. Surfaces that are odd-shaped or nonrectangular (triangular, circular, etc.) will result in more waste. Plywood is sold in 4′ × 8′ sheets. The builder must therefore calculate not only the total square footage but also the total number of plywood sheets to order.

Glued-laminated (also called glulam) wood is a product in which individual pieces of wood are bonded together with adhesives and then compressed to make a single piece. One of its biggest advantages is that it can be produced in almost any shape or size, offering practical and economical solutions for curved shapes and long spans. Glued-laminated sections are covered in RSMeans Section 06 18 13.

Wood pieces, natural or glued-laminated, are connected by direct means, such nails and screws, especially in small or thin pieces, or by using a wide variety of connectors. All wood fasteners are covered in Section 06 05 23.

Thermal and Moisture Control

Division 7 includes materials that prevent the penetration of water or heat through the building envelope, including roofing, siding, insulation, fireproofing, and sealants.

Most of these materials are measured by area: either square feet (SF) or squares (1 SQ = 100 SF). Some utilize units of CSF, which is the same as one square.[2] The area could be measured horizontally, as with roofs, or vertically, as with walls. Some liquid items are estimated by the gallon. The estimator has to know the conversion factor (e.g., SF/gallon).[3]

Roofs

The unit of measurement for roofing is the square (SQ). Exercise caution in estimating items in this division. Although sheathing, felt, and shingles occupy the same square foot area, sheathing is measured by the square foot unit (SF), and roofing materials and felt are measured by the square (SQ).

---

2. The letter *C* in the CSF acronym means 100 in Roman numerals. The letter *M* in the MBF (1,000 board feet) also means 1,000 in Roman numerals. The user has to be careful in interpreting the letter *M* because while it means 1,000 in some units, it means 1,000,000 in other uses, especially in electricity and electronics. In this case, it means the Greek word *mega*, which is a prefix meaning 1 million, as in megahertz, megapixel, and megabyte.

3. Again, the user has to be careful to distinguish the U.S. gallon, 3.785 liters, from the imperial (UK) gallon, 4.546 liters.

To calculate the area of one face of a sloped roof, first calculate the total length of the rafters (including any overhangs), and then multiply the rafter length times the length of the roof (again, including any overhangs). If the roof is a simple symmetrical gabled roof (with equal pitch and dimensions on both sides), then the total roof area is double the area of one face. For roofs with unequal slopes or areas, each face must be calculated separately.

Roof slope is usually expressed as a ratio of the roof's rise over run, or the number of inches of vertical change (in height) per 12 inches of horizontal change.

Total rafter length can be calculated using the Pythagorean theorem and the ratio for rise over run. The length of the rafter span is the distance from the face of the ridge to the exterior face of the building plus any roof overhang.

For a roof with an 8 in 12 slope, if the total run, B, is 13 feet, then the rise, A, $= 13' \times 8 / 12 = 8.667'$, and using the Pythagorean theorem, we find the rafter length, C, as shown in Figure 9.1:

$$A^2 + B^2 = C^2$$
$$(8.667)^2 + (13)^2 = C^2$$
$$244.1 = C^2$$
$$C = 15.6 = 15'-7''$$

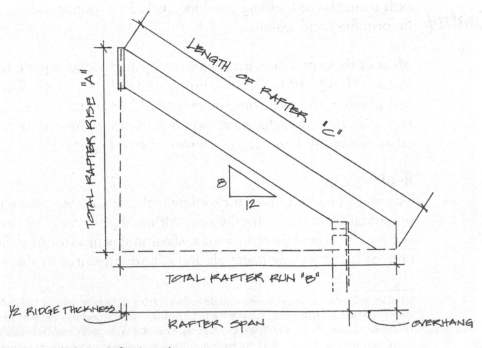

Figure 9.1 Calculating Rafter Length

Reprinted from *Builder's Essentials: Plan Reading and Material Takeoff*, by Wayne J. DelPico, RSMeans Drawing used with permission.

A simpler method for finding total roof area is to multiply the length of the roof by the horizontal projection times the roof factor (see Table 9.1).

The roof factor $= \sqrt{1+S^2}$ where S = the slope in inches per foot.

Table 9.1   Roof Factors

Rise	Run	Slope	Pitch	Roof Factor
1	12	0.0833	0.0417	1.0035
2	12	0.1667	0.0833	1.0138
3	12	0.2500	0.1250	1.0308
4	12	0.3333	0.1667	1.0541
5	12	0.4167	0.2083	1.0833
6	12	0.5000	0.2500	1.1180
7	12	0.5833	0.2917	1.1577
8	12	0.6667	0.3333	1.2019
9	12	0.7500	0.3750	1.2500
10	12	0.8333	0.4167	1.3017
11	12	0.9167	0.4583	1.3566
12	12	1.0000	0.5000	1.4142

Then

$$\text{Roof area} = \text{Rafter length} \times 2 \text{ sides} \times \text{Length of building}$$

or

$$\text{Roof factor} \times \text{Building horizontal area (including any projections)}$$

Example 1

Estimate the total floor and roof areas for the building shown in Figure 9.2. Also find the rafter count, assuming 16-inch spacing.

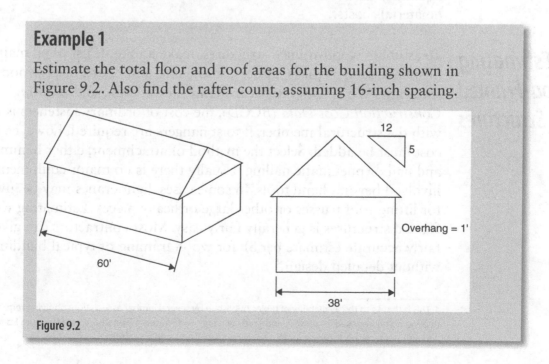

Figure 9.2

171

Solution

$$\text{Building's floor area} = \text{Length} \times \text{Width}$$
$$= 60' \times 38' = 2{,}280 \text{ SF}$$

For the roof (Figure 9.2), we can find the area in two ways:

The roof factor (from the table) = 1.0833
The roof horizontal projection area = $60' \times 40' = 2{,}400$ SF (with 1′ horizontal projection on each side of the width; there are no projections along the length.)
Roof area = $2{,}400$ SF $\times 1.0833 = 2{,}600$ SF

or

Rafter length = $20'$ ($\frac{1}{2}$ the width of the bldg. + 1′ horiz. projection) $\times 1.0833$
$= 21.67'$ (use 22′ long rafter)

Roof area = $21.67' \times 60'$ long $\times 2$ sides $= 2{,}600$ SF
No. of rafters on each side of the bldg $= 60'/(16''/12 \text{ in}/\text{ft}) + 1$

or

$$= (60'/1.33') + 1 = 46$$
Total number of rafters = 46×2 sides = 92 pieces, 22′ long each

For estimating sheathing, the quantity is 2,600 SF. It is common practice to estimate sheathing in number of sheets. In 4×8 sheets, this means $2{,}600 / 32 = 82$ sheets. A typical contractor will order a few more sheets than calculated. For felt and shingles (or any other roofing material), the quantity used is 26 SQ. A waste allowance then has to be added to all materials costs.

Estimating Wood-Framed Structures

To estimate wood-framed structures, make a takeoff list of all main members, such as joists, girders, studs, sills, and board or plywood sheathing. When using RSMeans Online estimating and *Building Construction Cost Data* (BCCD), the cost of ordinary fasteners is included with the structural member. If joist hangers are required, however, their cost must be added. Select the method of attachment, either hammer and nail or pneumatic nailing. Usually there is no major equipment involved beyond hand tools. In some cases, light cranes may be involved for lifting roof trusses or other large or heavy pieces. Estimating wood-framed structures is generally fairly easy. Most contractors can give a fairly accurate estimate per SF for wood framing of typical buildings, even without detailed design.[4]

---

4. This is thanks to the availability of tables published by several lumber associations showing required sections depending on imposed load, span, and grade of wood. See Southern Pine Council's "Maximum Spans" for joists and rafters, for example (www.southernpine.com/span-tables/joists-rafters).

Example 2

Estimate the total bare cost for furnishing and installing the next items:

a. Three 2×12 built-up girders

b. 2×10 floor joists

c. Cross-bridging

d. ¾″ plywood subfloor sheathing

Assume the house is to be built in Duluth, Minnesota (Figures 9.3 and 9.4).

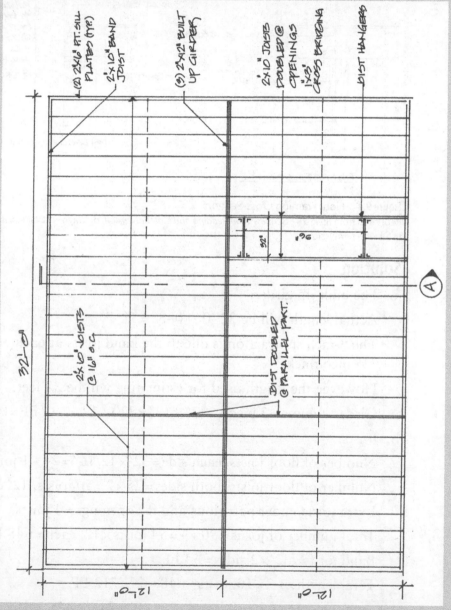

Figure 9.3   Floor Framing Plan

Reprinted from *Builder's Essentials: Plan Reading and Material Takeoff,* by Wayne J. DelPico, RSMeans.
Drawing used with permission.

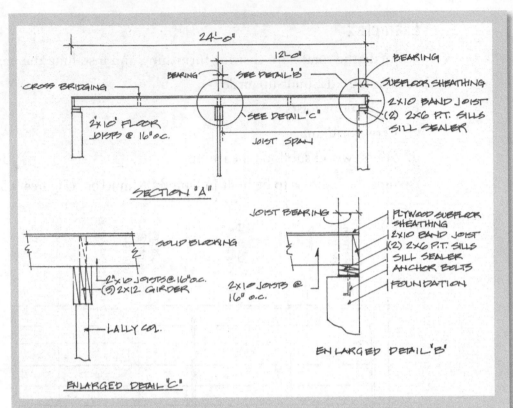

Figure 9.4 Floor Framing Cross-Section

Reprinted from *Builder's Essentials: Plan Reading and Material Takeoff*, by Wayne J. DelPico, RSMeans. Drawing used with permission.

Solution

a. The built-up girder:

Actual length will be 32'-0" – 0'-3" = 31'-9".

The 3-inch subtraction is due to the band joists at each end. Each is 1½-inch thick.

However, the length used for estimating will be 32 feet.

Girder volume = 3 pieces × (2" × 12" × 32') / 12 = 192 BF = 0.192 MBF

b. Joists:

Number of floor joists, each side = 32' × 12/16" = 24 + 1 joist = 25 joists

Number of floor joists, both sides = 25 × 2 = 50 joists, 12' each

Extra joists under partitions and for openings = 4 joists

Total number of joists = 50 + 4 = 54 joists × 12' each = 648 LF

Band joists = 32' × 2 sides = 64 LF

Total length of 2 × 10 joists = 648 + 64 = 712 LF

Total volume of 2 × 10 joists = 2" × 10" × 712' / 12 = 1,187 BF = 1.187 MBF

Note: RSMeans costs are based on the assumption that joists are spaced at 16″ OC. If the joist spacing is different (e.g., 24″ OC), the cost can be approximated by dividing by an adjustment factor of 24 / 16 = 1.5.

c. Cross-bridging:

The plan shows one cross-bridging pair at midspan between every two joists. Since there are 25 joists on each side, there must be 24 bridging pairs on each side. RSMeans Online estimating and *BCCD* list 1″ × 3″ cross-bridging cost per Pr (pair), with the option of manual or pneumatic nailing. In our case, assume manual.

$$\text{Total number of bridging pairs} = 24 \times 2 = 48 \text{ pairs}$$

The cost of bridging is based on joist depth, which is not specified in RSMeans Online estimating or *BCCD*. Joists are most frequently either 2 × 10 or 2 × 12, and the cost difference will be insignificant.

d. Plywood subfloor sheathing:

$$\text{Total floor area} = 32' \times 24' = 768 \text{ SF}$$

This building's dimensions are optimum for minimum waste. Not only do the joists have no design waste, but plywood subfloor sheathing has no waste either. Exactly 24 sheets (4′ × 8′) of plywood are needed. The table shows the unit cost for each item.

RS Means Item No.	Description	Quant.	Unit	Bare costs				Total Incl. O&P
				Mat.	Labor	Equip.	Total	
06 11 10.10 5060	3 – 2 × 12 girder	0.192	MBF	845.25	266.34	–	1111.59	1342.32
06 11 10.18 2720	2 × 10 joists	1.187	MBF	796.95	511.45	–	1,308.40	1655.88
06 11 10.06 0012	1 × 3 bridging	48	Pr	0.66	2.92	–	3.58	5.19
06 16 23.10 0200	¾″ subfloor plywood CDX	768	SF	0.91	0.61	–	1.52	1.93

e. Total bare cost:

$$\text{Total bare cost for the built-up girder} = 0.192 \times \$1,111.59$$
$$= \$213.43$$

Other items are calculated similarly.

$$\text{Total bare cost} = \$213.43 + \$1,553.07 + \$171.84 + \$1,167.36 = \$3,106 \approx \$3,100$$

Example 3

Estimate the total cost including overhead and profit for the job in Example 2.

Solution

Method 1: Using RSMeans total including O&P:

Total cost including O&P = 0.192 MBF × $1,342.32/MBF + 1.187 MBF × $1,655.88/MBF + 48 Pr × $5.19/Pr. + 768 SF × $1.93/SF = $3,955 ≈ $4,000

Method 2: Using RSMeans Total bare cost + Labor add-ons + 10% of all expenses for profit:

Labor add-ons:

Workers' comp. for carpentry—3 stories or less, Minnesota	= 13.9% × 156% = 21.68%
Average fixed overhead	= 18.3%
Overhead	= 11.00%
Profit	= 10.00%
Total labor add-ons	= 60.98%

Total cost:

Girder:

Materials = 0.192 MBF × $845.25/MBF	= $162.29
Materials profit = $162.29 × 10%	= $ 16.23
Labor = 0.192 MBF × $266.34/MBF	= $ 51.14
Labor add-ons = $51.14 × 60.98%	= $ 31.18
Total for girder	= $260.84

Repeating the same steps for the joists, bridging, and plywood, the totals are: $2,017.87, $260.48, and $1,522.93 for the joists, bridging, and plywood, respectively.

Total cost including O&P = $4,062

There is a difference of $107, which is less than 3 percent due to reasons explained earlier.

Example 4

Estimate the bare cost and cost including overhead and profit for providing and installing studs and plates for the exterior walls of Example 2. The studs will be 2 × 6 spaced at 16″ OC. There will be two top and one bottom 2 × 6 plates. Assume that the total number of studs is 106. The net length of the stud is 7′-6″.

Solution

Although the net length of the stud is 7′-6″, we will use the length 8 feet in the calculations since studs are sold in length increments of 2 feet. The perimeter of the building is $2 \times (24' + 32') = 112$ LF. If we use the same length of pieces for plates, we need $3 \times (112' / 8) = 42$ pieces.

RSMeans Item No.	Description	Unit	Bare Costs				Total Incl. O&P
			Mat.	Labor	Equip.	Total	
06 11 10.40 6040	2×6 plates	MBF	608.58	1013.25	–	1,621.83	2,210.54
06 11 10.40 6160	2×6 studs	MBF	608.58	762.35	–	1,370.93	1,824.54

Quantities:

Plates = $42 \times 2'' \times 6'' \times 8'/12 = 336$ BF = 0.336 MBF

Studs = $106 \times 2'' \times 6'' \times 8'/12 = 848$ BF = 0.848 MBF

Total quantity to be ordered = 148 pieces, or 1.184 MBF plus a suitable allowance for waste

Total bare cost:

Plates: 0.336 MBF × $1,621.83   = $ 544.93

Studs: 0.848 MBF × $1,370.93   = $1,162.55

Total bare cost   = $ 1,707

Total cost including O&P =

0.336 MBF × $2,210.54 + 0.848 MBF × $1,824.54 = $ 2,290

Example 5

Estimate the bare cost and cost including overhead and profit for providing and installing ceiling joists, rafters, sheathing, #15 asphalt felt, and asphalt shingles for the building in Example 2. The ceiling joists will be 2×12 and the rafters 2×8, both spaced at 16″ OC. Sheathing is ½-inch-thick CDX plywood. Shingles are asphalt, standard laminated class A. The roof has a 3 in 12 slope and will extend 1′ on all sides, measured horizontally.

Solution

The building's outside dimensions are 32′×24′. The roof's horizontal projection is 34′×26′.

Ceiling joists: 2 sides × [32′ × 12/16 + 1] = 50 joists × 12′ each = 600 LF

Joists quantity (volume) = 600′ × 2″ × 12″ / 12 = 1,200 BF = 1.2 MBF

Slope = 3:12 Roof factor = 1.0308

Rafter length = 13′ × 1.0308 = 13.4′. Use 14′ length.

Number of rafters = 54, Total length = 54 × 14′ = 756 LF

Rafters quantity (volume) = 756 LF × 2″ × 8″/12 = 1,008 BF = 1.008 MBF

Roof area = 34′ × 26′ × 1.0308 = 911.2 SF = 9.11 SQ

RSMeans Item No.	Description	Unit	Bare Costs				Total Incl. O&P
			Mat.	Labor	Equip.	Total	
06 11 10.18 2740	Joists 2×12	MBF	$845.25	$434.25	$0.00	$1,279.50	$1,598.04
06 11 10.30 7060	Rafters 2×8	MBF	$632.73	$603.13	$0.00	$1,235.86	$1,621.92
06 16 36.10 0100	Sheathing CDX ½″	SF	$0.61	$0.54	$0.00	$1.15	$1.50
07 51 13.10 0200	#15 felt asphalt coated	SQ	$5.69	$6.32	$0.00	$12.01	$17.01
07 31 13.10 0300	Asphalt shingles Std lam. Class A	SQ	$115.65	$79.56	$0.00	$195.21	$262.52

Bare cost for joists = 1.2 MBF × $1,279.50   = $1,535.40

Bare cost for rafters = 1.008 MBF × $1,235.86 = $1,245.75

Bare cost for sheathing = 911 SF × $1.15   = $1,047.65

Bare cost for felt = 9.11 SQ × $12.01   = $ 109.41

Bare cost for shingles = 9.11 SQ × $195.21   = $1,778.36

Total   = $5,717

Total cost incl. O&P = 1.2 MBF × $1,598.04 + 1.008 MBF × $1,621.92 + 911 SF × $1.50 + 9.11 SQ × ($17.01 + $262.52) = $7,466

Using Units of Quantity

When you calculate items in Divisions 6 and 7, be careful to note the proper units. Items may be tabulated per piece (each), CLF (100 lineal feet), MLF (1,000 lineal feet), VLF (vertical lineal feet), MBF (1,000 board feet), or other units. Roof area is calculated per SF (square foot) for items in Division 6 such as sheathing, and in SQ (square = 100 square feet) for other items in Division 7. Some items, such as trusses and frames, are given in dollars per SF Flr (area of the floor in square feet) as compared to some items in SF Roof (area of the roof in square feet).

Example 6

A building with a roof area of 8,750 SF in Indianapolis, Indiana, needs to be roofed. Assume a 160-mil SBS modified bitumen roofing.

Solution

Use line item 07 52 16.10 0750:

RSMeans Item No.	Description	Unit	Bare Costs				Total Incl. O&P
			Mat.	Labor	Equip.	Total	
07 52 16.10 0750	SBS modified bitumen roofing, 150 to 160 mils	SF	1.87	0.97	0.22	3.06	3.95

$$\text{Bare cost} = 8{,}750 \text{ SF} \times \$3.06 \qquad = \$26{,}775$$
$$\approx \$26{,}800$$
$$\text{Total cost incl. O\&P} = 8{,}750 \text{ SF} \times \$3.95 \qquad = \$34{,}563$$
$$\approx \$34{,}600$$

Chapter 9 Exercises—Set A

For solutions to the exercises in Set A, see the link to the student companion website at www.wiley.com/go/constructionestim5e.

1. What is the total volume, in MBF, for 40 pieces, 2 × 8, 16 feet long?

For the floor plan shown in Figure 9.6, calculate the total bare cost and cost including O&P for furnishing and installing the items in Exercises 2 through 9. Assume pneumatic nailing and the location to be Oshkosh, Wisconsin.

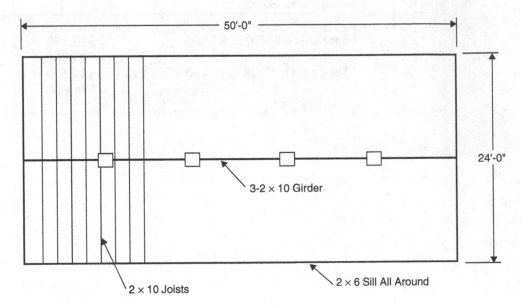

Figure 9.6

2. Subfloor girder, use three 2 × 10s.

3. Floor sills, use 2 × 6s.

4. Floor joists (between the 50′-long wall and the girder), use 2 × 10s, spaced at 16″ OC, including joist headers. Add seven extra joists under partitions.

5. Steel galvanized bridging.

6. Wall studs (for exterior walls only), use 2 × 4s, spaced at 16″ OC, including plates (2 top and 1 bottom), double studs at each of 12 openings, three (two extra) studs for each corner, and two extra studs at the nine intersections of partitions with the exterior wall.

   Note: For Exercises 7 through 9, assume that the roof extends 1 foot—measured horizontally—from all sides, and the slope is 4:12.

7. Rafters, use 2 × 8s, spaced at 16″ OC. Include a 2 × 10 ridge board.

8. Roof sheathing, use ½"-thick CDX plywood boards.

9. Asphalt felt #15 and asphalt shingles, standard laminated, class A.

10. Calculate the total bare cost for all items in Exercises 2 through 9.

11. Calculate cost including O&P for all items in Exercises 2 through 9.

12. Redo Exercises 7, 8, and 9 assuming a slope of 7:12.

13. Redo Exercise 10 if the carpenter is paid $38/hour.

14. Redo Exercises 2 through 11 if labor productivity is down 25 percent due to severe cold in Wisconsin.

15. Redo Exercises 2 through 11 if workers are working 12 hours per day, 6 days per week. Assume a 2-week duration, with a payroll cost factor of 1.5.

Chapter 9 Exercises—Set B

Solutions to Set B exercises are available to instructors only; see the link to the instructor's companion website at www.wiley.com/go/constructionestim5e.

1. What is the total volume, in MBF, for 300 studs, 2 × 4, 8′ long?

   For the floor plan in Figure 9.7, calculate the total bare cost and cost including O&P for furnishing and installing the items in Exercises 2 through 9. Assume the location to be Altoona, Pennsylvania.

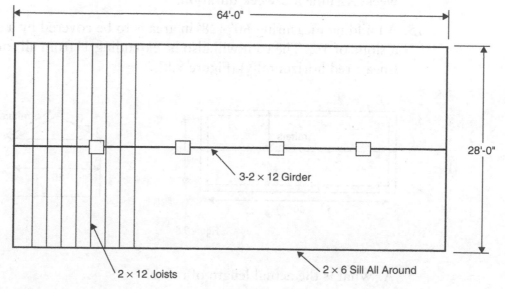

Figure 9.7

2. Subfloor girder, use three 2×12s (see Figure 9.7).

3. Floor sills, use 2×6s.

4. Floor joists (between the 64'-long wall and the girder), use 2×12s, spaced at 16" OC, including joist headers. Add 11 extra joists under partitions.

5. Steel galvanized bridging.

6. Wall studs (for exterior walls only), use 2×4s, spaced at 16" OC, including plates (two top and one bottom) and headers and double studs for openings. Also use three (two extra) studs for each corner.

   For questions 1 through 14, assume that the roof extends 1', measured horizontally, from all sides, and the slope is 3:12.

7. Rafters, use 2×8s, spaced at 16" OC. Include a 2×10 ridge board.

8. Roof sheathing, use ⅝"-thick CDX plywood boards.

9. Asphalt felt #30 and asphalt shingles, premium laminated, class C.

10. Calculate the total cost and cost including O&P for all items in Exercises 2 to 9.

11. Redo Exercises 7, 8, and 9, assuming a slope of 7:12.

12. Redo Exercise 4 if the carpenter is paid $40/hour.

13. Redo Exercise 4 if labor productivity is down 20 percent due to severe cold in Pennsylvania.

14. Redo Exercise 4 if workers are working 10 hours per day, 6 days per week. Assume a 2-week duration.

15. A building measuring 60'×28' in area is to be covered by a roof with a slope of 1:3. The roof will also be extending 1' from all four sides (measured horizontally) (Figure 9.8).

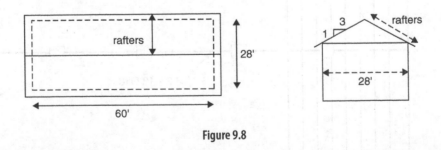

Figure 9.8

A. What is the actual length of rafter?

B. What is the length of the rafter to be ordered?

C. If rafters are spaced at 16″ on center, how many rafters do we need? (Rafters extend to the full width as shown, but there are no rafters at the end of the longitudinal extension.)

D. Including reasonable waste allowance, how many rafters to be ordered? What is the total volume (BF or MBF) of rafters order assuming 2 × 8 size?

E. Estimate the bare cost for providing and installing the rafters if the building is located in Las Vegas, Nevada.

Chapter 10

Doors and Windows, Interior Finish, and Equipment
Divisions 8 to 14

Using RSMeans Costs

The methodology for using RSMeans costs for items in Divisions 3 to 7 has been covered in previous chapters. The same principles, for the most part, apply to cost items in other divisions. The differences are largely related to quantity takeoff concepts, which are not the main object of this text.

Solving Estimating Problems

Cost estimating scenarios using RSMeans costs usually involve one or more of these tasks:

1. Estimating the total bare (direct) cost and the bare cost per unit (e.g., square foot)

2. Adjusting prices to reflect project costs in a particular city in the United States or Canada (choosing the location in the Online Estimating)

3. Estimating the total cost including overhead and profit using RSMeans total costs including overhead and profit

4. Estimating the total cost, including overhead and profit, using RSMeans total bare costs, plus labor burden and a percentage of bare costs for overhead and profit

5. Interpolating and extrapolating between line items, if feasible, to find the cost of a new work item that differs in dimensions, or other parameters, from similar items in RSMeans data

185

6. Adjusting a cost estimate to reflect productivity changes due to weather or other factors

7. Estimating the duration of a job

8. Evaluating the effect of overtime on production to be included in the cost estimate

9. Adjusting the cost estimate for using a contractor's local wages instead of the RSMeans labor rates

10. Adjusting a cost estimate for using equipment owned or rented for either a short term or an extended period of time

11. Adjusting a cost estimate to reflect anticipated inflation (or escalation of prices) for projects to be built in the near future

12. Using, whenever possible, RSMeans reference information

Division 8: Openings

This division covers openings, such as:

- Passage doors and frames
- Specialty doors
- Entrances
- Storefronts
- Curtain walls
- Windows
- Roof windows and skylights

Hardware, glazing, and louvers and vents are also included. Door and window openings are rough-framed by the work crew erecting the wall (carpenter, mason, etc.). The window frames, sash, and glazing are typically installed by a specialty subcontractor. When pricing doors, the estimator should use care to find out whether the door is prehung. If not, the cost of a frame must be added. The cost of hardware items, including passage sets, bolts, deadlocks, closers, panic devices, weather stripping, and so forth, should be added to all doors and windows. Painting doors has to be added, in most cases.

The cost of residential windows typically includes glazing. For commercial windows, the cost of the sash may not include glazing. Windows, doors, and panels in commercial buildings are usually installed without glass. The glazing is performed as a separate step and appears as a separate cost item in the estimate. A glazing takeoff is usually done by the square foot. Occasionally glazing will be quantified using the unit "united inches." United inches represent the sum of glass width plus height in inches. A glass panel 8″×20″ would be 28 united inches.

When estimating the items in the openings section, make sure the type of door, window, or hardware found in the *RSMeans Building Construction Cost Data* (*BCCD*) or quoted by a vendor matches what has been specified. Differences that look minor may have a significant impact on the cost.

Division 9: Finishes

This division addresses all items used for interior finish of walls, floors, and ceilings. The major subsections of Division 9 are for plaster and gypsum board, tiling, ceilings, flooring, wall finishes, acoustic treatments, and painting and coating.

The plaster and gypsum board section includes the cost of metal studs, lath, plaster, and gypsum board products. Lath is a base for plaster, consisting of wood strips, expanded metal mesh, or gypsum panels. In current construction, almost all lath used is metal mesh, or *wire lath*. Furring consists of strips of wood or metal fastened to a wall or ceiling in order to provide an even surface for the application of an interior finish, such as plaster. Furring may also be used to create an air space or to give the appearance of greater thickness. Lath can be used with or without furring. The takeoff unit for lath and furring is square feet, square yards, or linear feet.

Gypsum board is commonly used for interior walls and partitions. It is typically purchased in 4′ × 8′ or 4′ × 12′ sheets. Openings are not usually deducted unless they extend from floor to ceiling. To estimate the quantity required, multiply the perimeter of the area by the height to obtain a total area. Divide by the area of the sheet to determine the number of sheets required. Gypsum board requires finishing of the seams with tape and compound before wall finish is applied. The five levels of interior finish recognized by the construction industry are defined in R092910–10. The most common is level 4, used for walls where wall covering is applied or a flat paint is used.

Tiles include ceramic tiles, quarry tiles, structural clay tiles, vinyl composition tile, and a wide range of other materials. Tiles are available in a wide range of sizes, shapes, and colors. They may be used as a finish for floors or walls. Tile may be embedded in concrete (mud-set) or adhered with mastic (thin-set).

Ceilings, Flooring, and Wall Finishes

Ceilings may consist of tiles or boards adhered to framing and furring or suspended on a grid, and are priced by the square foot. Flooring materials include wood products, resilient products, and carpet. Wood and resilient floors are priced by the square foot. Carpet is priced by the square yard.

When estimating flooring materials that are purchased in sheets, particular care should be given to anticipated waste. The amount of waste will depend on the dimensions of the space, the roll width, and the repeated pattern of the flooring. A carpet plan may be provided with the contract drawings specifying a layout of carpet materials. Waste in carpeting may range between 3 and 30 percent. The waste factor is applied to the material only.

Wall finishes include wall coverings and paneling. When calculating quantities of wall covering, multiply the length of the wall by the height, and delete the area of all openings over 4 square feet (SF). The actual quantity purchased will depend on pattern match. Waste can be as much as 25 to 30 percent. The waste factor is applied to materials only. Most wall coverings are priced by the square foot.

Acoustic treatment includes items that limit the movement of sound. Items in this category include blankets and foam products, plenum barriers, and sound-absorbing wall panels. The unit of takeoff is usually square feet.

Paint and coatings include all the work items required to complete interior and exterior painting. The majority of the cost is for labor, so a careful listing of all tasks is imperative. Prep work includes cleaning, sanding, filling, and masking surfaces. The estimate should include the type of paint to be used (e.g., oil or latex), the method of application (brush, roller, or spray), and the number of coats.

The area to be painted is usually calculated in square feet by multiplying length by height (or width). Deductions are made for openings greater than 4 SF. When estimating a job that involves roller or spray work, be sure to add the cost of all cut-ins using brushwork. When estimating cut-ins or trim work, calculate the area by using the factor 1 LF = 1 SF.

Example 1

Estimate the cost of surface preparation before painting an office area in a new building. The room is 50′ × 85′, with a ceiling height of 14 feet. The area contains six flush metal doors, 3′ × 7′ each, and 24 casement windows, 3′ × 5′ each. The project is located in Seattle, Washington.

Solution

Floor area: $50' \times 85' = 4,250$ SF

Wall area: $[2 \times (50' + 85') \times 14'] - (3' \times 5' \times 24) = 3,420$ SF. Cut-ins are the surfaces where the roller cannot reach or cover well. They have to be painted by brush. Cut-ins are measured in linear feet. In our examples, they represent:

- The meeting lines between the walls and the floor: $2 \times (50' + 85')$
- The meeting lines between the walls and the ceiling: $2 \times (50' + 85')$
- The inside corners where walls intersect: $4 \times 14'$
- The edges around windows: $24 \times (3' + 5' + 3' + 5')$
- The edges between the doorframe and the wall: $6 \times (7' + 3' + 7')$

Total cut-ins: $4 (50' + 85') + (4 \times 14') + (24 \text{ windows} \times 16') + (6 \text{ doors} \times 17') = 1,082$ LF

RSMeans Item No.	Description	Unit	Mat.	Labor	Equip.	Total	Total Incl. O&P
09 91 03.40 0150	Hand wash metal door, flush	SF	–	.15	–	.15	.22
09 91 03.40 0660	Interior walls, sand, gypsum board and plaster, light	SF	–	.11	–	.11	.15
09 91 03.20 0100	Sand and putty int. trim as comp. to paint 1 coat, quality work	LF	–	100%	–	–	–
09 91 23.52 7250	Paint trim, wood, under 6" wide, 1 coat, brush	LF	.05	.49	–	.54	.81
09 91 03.20 0530	Volume cover-up, plastic sheeting	SF	–	.02	–	.02	.03

The cost for surface preparation involves inspecting and lightly sanding the walls, sanding and filling the woodwork and trim, and applying plastic sheeting to cover the floor area.

Item 09 91 23.52 7250 in the preceding table is not used in the estimate. It is listed only as a reference to item 09 91 03.20 0100, which is a percentage of the earlier item.

To calculate the bare cost:

Doors: 2 sides × 6 door × 21 SF × $0.15	= $ 37.80
Walls: 3,420 SF × $0.11	= $376.20
Trim: 1,082 × $0.49	= $530.18
Volume cover-up (floor): 4,250 SF × $0.02	= <u>$ 85.00</u>
Surface preparation total:	= $1,029

The total with overhead and profit is:

$$(252\,SF \times \$0.22) + (3{,}420\,SF \times \$0.15) + \left(1{,}082\,LF \times \$0.755^{[1]}\right)$$
$$+ (4{,}520\,SF \times \$0.03) = \$1{,}521$$

All unit prices are already adjusted to the location: Seattle, Washington.

Example 2

Estimate the painting cost for the above project. Use latex paint on the walls and oil-based (alkyd) paint on the woodwork and doors.

Item	RSMeans Item No.	Description	Unit	Mat.	Labor	Equip.	Total	Total Incl. O&P
1	09 91 23.72 0280	Paint conc. or drywall, alkyd primer, spray	SF	0.06	.12	–	0.18	.23
2	09 91 23.72 0880	Paint 2 coats alkyd finish, spray	SF	0.13	.20	–	0.33	.44
3	09 91 23.52 7200	Paints trim, primer coat, oil base, brushwork, under 6" wide, incl. puttying	LF	0.03	.49	–	0.52	.79
4	09 91 23.52 7450	Paint trim, under 6", oil base, 3 coats, brushwork	LF	0.15	1.00	–	1.15	1.66
5	09 91 23.33 0500	Paint doors and frame, 3'×7' primer	Ea	3.81	32.43	–	36.24	53.07
6	09 91 23.33 1000	Paint doors and frame, 3'×7', alkyd finish	Ea	3.99	32.43	–	36.42	53.28
7	09 91 23.33 8600	Paint window incl. frame and trim, 1 side, 3'×5' opening, alkyd primer	Ea	0.99	16.17	–	17.16	25.28
8	09 91 23.33 8800	Paint window incl. frame and trim, 1 side, 3'×5' opening, alkyd finish	Ea	1.58	16.17	–	17.75	25.93

---

1. This number represents the number shown in the table, $0.81, minus the materials part including its 10 percent profit; $0.05 × 1.1.

Solution

First, calculate the painting of the walls. The bare cost is:

Wall, prime:
$3,420 \text{ SF} \times \$0.18$ = $ 615.60

Wall, finish:
$3,420 \text{ SF} \times \$0.33$ = $1,128.60

Cut-ins, prime:
$1,106 \text{ LF} \times \$0.52$ = $ 575.12

Cut-ins, finish:
$1,106 \text{ LF} \times \$1.15$ = $1,271.90

Paint walls total: = $3,591.22

The cost with overhead and profit is:

$$3,420 \text{ SF} \times (\$0.23 + \$0.44) + 1,106 \text{ SF} \times (\$0.79 + \$1.66) = \$5,001$$

The bare cost for painting the doors and windows is:

Doors, prime:
$6 \times \$36.24$ = $ 217.44

Doors, finish:
$6 \times 2 \text{ coats} \times \$36.42$ = $ 437.04

Windows, prime:
$24 \times \$17.16$ = $ 411.84

Windows, finish:
$24 \times 2 \text{ coats} \times \$17.75$ = $ 852.00

Paint doors and windows total: = $1,918.32

The cost with overhead and profit is:

$$6 \times (\$53.07 + 2 \times \$53.28) + 24 \times (\$25.28 + 2 \times \$25.93) = \$2,809$$

The total bare cost of surface prep, priming, and painting the office area is (rounded):

$$\$1,029 \, (\text{from Example 1}) + \$3,591.22 + \$1,918.32 = \$6,539$$

And the total cost of surface prep, priming, and painting the office area, including overhead and profit, is:

$$\$1521 + \$5,001 + \$2,809 = \$9,331$$

The results are tabulated in an Excel file produced by the RSMeans Online Estimating, as shown in Table 10.1.

Table 10.1 Unit Cost Estimate

Quantity	LineNumber	Description	Unit	Ext. Mat.	Ext. Labor	Ext. Equip.	Ext. Total	Ext. Total O&P
252	099103400150	Surface preparation, interior, doors, hand wash, metal, flush	S.F.	$ -	$ 37.80	$ -	$ 37.80	$ 55.44
3420	099103400660	Surface preparation, interior, walls, sand, gypsum board and plaster, light	S.F.	$ -	$ 376.20	$ -	$ 376.20	$ 513.00
1082	099123527250	Paints & coatings, miscellaneous interior, trim, wood, paint 1 coat, oil base, brushwork, under 6" wide, incl. puttying	L.F.	$ -	$ -	$ -	$ -	$ -
1	099123527250	Paint preparation, sanding & puttying interior trim, compared to painting 1 coat, on quality work	L.F.	$ -	$ 530.18	$ -	$ 530.18	$ 811.50
4250	099103200530	Paint preparation, surface protection, placement & removal, volume cover up (using plastic sheathing or building paper)	S.F.	$ -	$ 85.00	$ -	$ 85.00	$ 127.50
				Total Example 10.1			**$ 1,029**	**$ 1,507**
3420	099123720280	Paints & coatings, walls & ceilings, interior, concrete, drywall or plaster, latex paint, primer or sealer coat, smooth finish, spray	S.F.	$ 205.20	$ 410.40	$ -	$ 615.60	$ 786.60
3420	099123720880	Paints & coatings, walls & ceilings, interior, concrete, drywall or plaster, latex paint, 2 coats, smooth finish, spray	S.F.	$ 444.60	$ 684.00	$ -	$1,128.60	$ 1,504.80
1106	099123527200	Paints & coatings, miscellaneous interior, trim, wood, primer coat, oil base, brushwork, under 6" wide, incl. puttying	L.F.	$ 33.18	$ 541.94	$ -	$ 575.12	$ 873.74

Quantity	LineNumber	Description	Unit	Ext. Mat.	Ext. Labor	Ext. Equip.	Ext. Total	Ext. Total O&P
1106	099123527450	Paints & coatings, miscellaneous interior, trim, wood, paint 3 coats, oil base, brushwork, under 6" wide	L.F.	$ 165.90	$1,106.00	$ -	$1,271.90	$ 1,835.96
6	099123330500	Paints & coatings, interior, alkyd (oil base), flush door w/frame, primer, brushwork, 3' x 7'	Ea.	$ 22.86	$ 194.58	$ -	$ 217.44	$ 318.42
12	099123331000	Paints & coatings, interior, alkyd (oil base), flush door w/frame, 1 coat, brushwork, 3' x 7'	Ea.	$ 47.88	$ 389.16	$ -	$ 437.04	$ 639.36
24	099123338600	Paints & coatings, interior, alkyd (oil base), windows, w/ frame & trim, per side, standard, single lite, oil, primer, brushwork, 3' x 5'	Ea.	$ 23.76	$ 388.08	$ -	$ 411.84	$ 606.72
48	099123338800	Paints & coatings, interior, alkyd (oil base), windows, w/ frame & trim, per side, standard, single lite, oil, paint 1 coat, brushwork, 3' x 5'	Ea.	$ 75.84	$ 776.16	$ -	$ 852.00	$ 1,244.64
				Total Example 10.2			**$5,510**	**$7,810**
				Total Examples 10.1 & 10.2			**$6,539**	**$9,318**

Examples 1 and 2 underscore the point that sometimes it takes several cost items to complete a work item. Had we included only the cost of spray painting the walls (i.e., second item in the table), the cost would have been only:

$$\text{Bare cost: } 3,420 \text{ SF} \times \$0.33 = \$1,129,$$

or

$$\text{Total cost including O\&P: } 3,420 \text{ SF} \times \$0.44 = \$1,505$$

Our error would have been $9,331/$1,505 = 6.2, which is more than 600 percent error.

Cost Modifiers

Many sections in the RSMeans database include cost modifiers. It is critical when using RSMeans data to become familiar with modifier lines and use them to fine-tune the estimate. There are modifiers for working at heights over 8 feet in the painting section, located at lines 09 91 13.60 8100 through 8300 and 09 91 23.72 8200 and 8300. In addition, there are adjustment factors for changing from alkyd to latex paint or latex to alkyd at the end of most sections.

Cost modifiers adjust the cost on a unit price line. If in Example 2 earlier we needed to use oil-based paint for the walls, we need to adjust the cost of the paint by adding 10 percent to the material cost (item 09 91 23.72 1700 or 4120). The process in RSMeans Online estimating is discussed next.

You already have the eight items selected earlier, and you want to apply a 10 percent increase to the materials cost only to the first two items (i.e., 09 91 23.72 0280 and 0880). After adding these two items to the estimate, add the item 09 91 23.72 4120. This is not a "stand-alone" item. It is an adjustment to other items. As soon as you add it, a screen opens with the two items in the estimate displayed (Figure 10.1). You need to select the item(s) to be adjusted by adding 10% to the cost of materials. Note that you cannot add more than one adjustment item at a time because different adjustment items may be linked to different items. The program will ask the user to link the adjustment item to an item(s). The adjustment item may be linked to several items, as you can see in Figure 10.1.

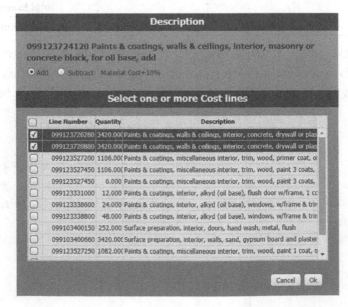

Figure 10.1

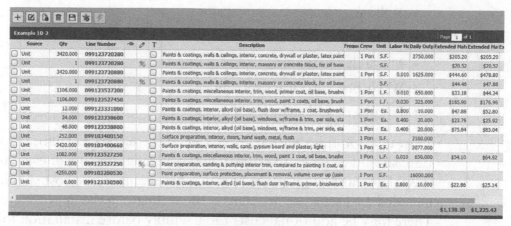

Figure 10.2

The original adjustment item number, 09 91 23.72 4120, disappears and each of the selected items appears twice: once as originally priced (with actual quantity), and once for the adjustment (with quantity showing as 1). This (quantity = 1) should not concern you as the adjustment is calculated automatically based on the actual quantity of the selected item.

Figure 10.2 shows the estimate with the addition for oil-based paint applied to it. Note the adjustment symbol +/ – in the customized column (title shows a pencil).

The notes field on the far right-hand side of the estimate allows the user to add notes for the adjustment. You can also access the note box by double-clicking the item and opening the "Line Item Details." These notes go with the item only in the estimate, not in the database. You cannot edit the items including the adjustment items.

Note: Each adjustment line has a specific purpose that is clear in the description of the line. These lines should be used only for the purpose for which they were intended. If none of the adjustment items in the database satisfies your purpose, create a custom item that matches the original item but adjusted as needed.

Division 10: Specialties

This division includes special items that enhance a building's utility and character including:

- Visual display boards
- Building signage
- Pedestrian control devices
- Toilet and shower compartments

- Toilet accessories
- Lockers
- Movable partitions
- Wardrobe and closet specialties
- Fireplaces and stoves

Most items in this division are priced by the units Ea (each), SF, or LF. Care should be given to estimating the installation cost of each item. If special anchors, support, or blocking are required, the cost should be included as a separate line item in the estimate in the appropriate division. There may be additional costs for assembly before installation for items such as lockers. Assembly can often exceed installation time.

Division 11: Equipment

The items in Division 11, Equipment, are categorized in these general groups:

- Vehicle and pedestrian
- Security, detention, and banking
- Commercial
- Residential
- Foodservice
- Educational and scientific
- Entertainment
- Athletic and recreational
- Healthcare
- Collection and disposal
- Other equipment

Usually equipment is specified and supplied by the owner and installed by the contractor. In such cases, as in Division 10, the contractor will need to provide support, anchorage, and blocking in the appropriate divisions of the estimate, and coordinate installation, including mechanical and electrical hookups. Most items are priced by the unit or square foot.

Division 12: Furnishings

Furnishings, the subject of Division 12, are usually priced by the unit. They are most often purchased by the owner and installed by the contractor. The vendor's price may or may not include delivery and installation. When installation is not included, it must be added to the estimate. If items require a support system or mechanical/electrical hookup, this must be included.

Division 13: Special Construction

The major subsections of Division 13 are:

- Special facility components
- Special-purpose rooms
- Special structures
- Integrated construction
- Special instrumentation

The special facility components section includes swimming pools, fountains, aquariums, amusement park structures and equipment, tubs and pools, ice rinks, and kennels and animal shelters. Special-purpose rooms include controlled-environment rooms, shelters and booths, planetariums, and special activity rooms, such as saunas and steam baths. Special structures include air-supported fabric structures, space frames, geodesic domes, pre-engineered buildings, and towers. Integrated construction covers prefabricated building modules and mezzanines; sound, vibration, and seismic control assemblies; and radiation protection. Special instrumentation includes instrumentation for measuring and recording phenomena such as stresses in structures, solar and wind energy, and the effects of earthquakes. The estimator must be sure to account for items needed for interface or integration between the special construction item and other parts of the building, such as the foundation.

Division 14: Conveying Equipment

Division 14 addresses the various types of mechanical means for transporting people, goods, or other objects from one location to another, either vertically or horizontally. This division includes:

- Dumbwaiters
- Elevators
- Escalators and moving walks
- Lifts
- Turntables
- Scaffolding
- Other equipment, including chutes and tube systems

Most of these systems are installed by the supplier or a specialized subcontractor. Some items require a type of support system not included in the system price. This additional support work must be included in the proper section of the estimate.

Chapter 10 Exercises—Set A

For solutions to the exercises in Set A, see the link to the student companion website at www.wiley.com/go/constructionestim5e.

1. Estimate the bare cost (materials, labor, and equipment) for erecting the frames for four steel doors using 8-inch channel at 11.5 lb/LF. Each door measures 6′ × 8′ and weighs 275 lb. Assume steel channels with anchors and bar stops for the framing. The job will be in Phoenix, Arizona.

2. Repeat Exercise 1 assuming frames without the bar stops.

3. Estimate the total cost for Exercise 1 including O&P using:

   A. RSMeans total incl. O&P.
   B. Calculate the total with overhead and profit using RSMeans bare cost plus labor add-ons and 10 percent for overhead and profit.

4. Estimate the cost of providing and installing 12 sliding windows, 5′ × 4′, including frame, screen, and grilles. Assume vinyl clad, premium with double insulating glass. The job will be in Miami, Florida.

5. Estimate the bare cost and cost including O&P for providing and installing metal toilet partitions for three cubicle units—two regular and one handicap. They are floor and ceiling anchored. Also add two urinal screens, 18 inches wide. All will be painted metal and will be erected in Dover, Delaware.

6. Assume the job in Exercise 5 is to be done by a subcontractor who will work under the general contractor. How much do you expect the general contractor to charge the owner for this job?

7. An owner contracted with a builder to build a home for him. He is thinking of adding a built-in 42″ hearth, radiant fireplace with small circulating fan. It needs a double-wall, stainless steel chimney 12″ diameter and 13′ high. The house is in Cleveland, Ohio. The owner is asking you (as a friend) how much this addition will cost so he can tell whether the $3,900 price tag the contractor put on it is reasonable.

8. What would an average cashier's booth for a parking garage cost in Minneapolis, Minnesota?

9. Estimate the bare cost and cost including O&P for the demolition of a 100′ × 60′ pre-engineered steel building with rigid frame and clear span in Ann Arbor, Michigan.

10. Repeat Exercise 9 if the demolition is needed for two identical and adjacent buildings of the size and location mentioned in Exercise 9.

11. Estimate the bare cost and cost including O&P for providing and installing a pre-engineered steel building in Jamaica, New York. The building is a single-post, two-span frame, 26-gauge, colored roofing and siding, 120′ long, 100′ wide, 20′ eave height. Add:

    A. 4′ eave overhang with soffit along the long sides

    B. 4′ × 8′ entrance canopy with frame

    C. Four doors, single leaf 3070 (3′ × 7′), economy

    D. 16 framed openings for windows 4030 (4′ × 3′)

    E. Gutter, eave type, galvanized, along the two long sides of the building

12. Estimate the bare cost and cost including O&P for providing and installing two passenger electric elevators for a six-story office building in Rapid City, South Dakota. The capacity is 4,000 lb and the speed is 300 fpm. There will be a total of six stops with a total vertical travel of 72 LF.

Chapter 10 Exercises—Set B

Solutions to Set B exercises are available to instructors only; see the link to the instructor's companion website at www.wiley.com/go/constructionestim5e.

1. An owner of a commercial building is replacing the front door that measures 3′-6″ × 8′. He is comparing these alternatives:

   A. Solid wood, decorator, pine

   B. Hand-carved mahogany

   Help the owner compare the cost of the two alternatives. Assume the job is in Evansville, Indiana. Use direct cost only for the comparison.

2. Estimate the cost of the frame in Exercise 1 using pine wood. Assume the frame is 5 3/16 ″ deep.

3. How much would a 16′ × 7′ deluxe metal garage door cost if it is to be constructed for a house in Berkeley, California:

   A. Bare cost (what it would cost the contractor)

   B. Total cost including O&P (what the contractor would charge the owner)

4. A subcontractor has to install drywall for walls and ceilings of a new apartment building. Assume the total area is 150,000 SF for walls and 48,000 SF for ceilings. Drywall for walls will be ⅝″ thick, taped and finished (level 4 finish). Drywall for ceilings will be ½″ thick (level 5 finish). The project is in Spokane, Washington. What is the bare cost and cost including O&P?

5. What would the answers of Exercise 4 be if the project will be built in 2019? Assume reasonable inflation/escalation factor.

6. Estimate the cost of painting the walls and ceilings in Exercise 4. Ceilings are painted with flat paint, two coats, sprayed. Walls are painted with oil-based paint, two coats, sand-finish sprayed. Do not include brushwork cut-ins.

7. Another subcontractor for the same project in Exercise 4 needs to provide the flooring:

   A. 5,000 SF 16″×16″ ceramic tile, glazed with thin set, color group 1

   B. 9,500 SF vinyl composite tile, 12″×12″, ³⁄₃₂″ thick, marbleized
   C. 7,500 SF carpet, wool, patterned, 48 oz, heavy traffic

   D. 24,000 SF carpet, nylon plush, 36 oz, medium traffic

   Estimate the bare cost and cost including O&P for the above items.

8. A middle school in Salem, Oregon, is looking for a contractor to build 32 metal lockers, double tier, 18″×18″×36″. How much do think the school would pay the contractor, and how much would it cost the contractor?

9. Estimate the bare cost and cost including O&P for providing and installing a drive-up window including a drawer and a microphone for a bank in Kansas City, Missouri. Add a 6′×12′ panel of 1.³⁄₁₆″ laminated bullet-resisting glass.

10. The apartment building in Exercise 4 need blinds for the windows and glass doors. The windows are:

    A. 270 windows 4′×4′

    B. 90 windows 4′×6′

    C. 90 windows 1′-8″×3′

    D. 60 double-glass doors, 8′×6′-8″

    Use 2-inch horizontal interior, aluminum stock (average) for all windows, and 4-inch vertical aluminum slats (average) for the glass doors. Estimate the bare cost and cost including O&P for this job.

Chapter 11

Fire Suppression, Plumbing, Mechanical, and Electrical
Divisions 21 to 28

The former MasterFormat 95 Divisions 15 and 16 (mechanical and electrical work) have been newly expanded and relocated in MasterFormat 2004 (with revisions up to April 2016) to Divisions 21 to 28. The major sections are:

- Division 21, Fire Suppression
- Division 22, Plumbing
- Division 23, Heating, Ventilating, and Air Conditioning
- Division 25, Integrated Automation
- Division 26, Electrical
- Division 27, Communications
- Division 28, Electronic Safety and Security

Normally these divisions are estimated by specialists well versed in these trades. The primary reason for this is that the plans and specifications show only the general configuration and layout of the systems. Many methods and materials utilized by these trades are standardized and understood by the installers, and are not usually detailed in the contract documents.

Takeoffs in these divisions require particular care. The level of detail requires a consistent methodology on the part of the estimator to ensure that nothing is missed. Here are a few tips for the estimator:

- Use preprinted forms. Always use forms with the same format and be consistent.
- Use only the front side of each sheet of paper.
- When copying numbers from one sheet to the next, transfer them carefully.
- Always list dimensions (width, length, and height) in a consistent order and don't forget to mention the units.
- Verify the scale of drawings before using them as a basis for measurement. Be on the lookout for changes in scale, photocopy reduction, or notes such as NTS (not to scale).
- Mark drawings neatly and consistently when counting quantities.
- Include required items that may not appear in the plans and specs.
- Be alert for discrepancies between the plans and specifications.
- Print legibly, and use pencil.
- Be organized and consistent.
- Use educated common sense.

The experienced estimator will visualize the work as he or she performs the quantity takeoff. It's a good practice to perform the estimate system by system. Think of the systems as a flow of work from service entrance, metering and control, distribution throughout the building, and end-use fixtures or appliances. Always refer to the specifications for definitions of quality, installation instructions, and any extras that must be provided, including extra parts, warranties, manuals, and training of building personnel.

Fire Suppression

Fire suppression systems within buildings include all systems to control the spread of fire. The major subdivisions are:

- Water-based fire suppression systems, including sprinklers and standpipes
- Fire extinguishing systems, including carbon dioxide (CO_2) and chemical extinguishers
- Fire pumps
- Fire suppression water storage tanks

Standpipe Systems

Standpipe systems are designed to deliver large quantities of water for fire control to remote areas of a building where laying hose by a fire department would be difficult or time-consuming. There are three general classes of service for standpipe systems, denoting whether they are for use of building or fire department personnel. Standpipe classes are defined in R211226–20, NFPA 14 Basic Standpipe Design.[1] These systems consist of Siamese connections, heavy pipe risers, valves, hose cabinets at each floor, hoses, hose racks, and alarms. Piping portions of these systems are taken off by the linear foot (LF). All other components and accessories are taken off by the unit each (Ea).

Sprinkler Systems

Sprinkler systems distribute water throughout the building for fire control. The most common type is the wet-pipe system, which is filled with pressurized water from a municipal system. When heat above a set temperature reaches a sprinkler's head, a fusible link in the head melts, activates immediately, and releases water to put out the fire. Quick response and low initial cost are the main reasons to install a wet-pipe system. If the area will be subject to below-freezing temperatures, dry-pipe systems are recommended. Several types of dry-pipe detection and actuation systems are in common use, providing the designer a choice. Sprinkler systems vary, depending on the classification of building occupancy by the level of hazard, which is determined by the building's contents and intended use. The system classification is shown in R211313–20.

Fire Pumps

Fire pumps are part of the fire suppression system within a building used to move water to upper floors at sufficient pressure to supply the sprinkler or standpipe system. The unit of takeoff is each (Ea).

Fire Suppression Water Storage Systems

Fire suppression water storage systems include underground, ground, and elevated storage tanks that contain water for fire control.

Plumbing

The plumbing system includes the piping and fixtures for the potable water, domestic hot water, storm and sanitary waste, and interior gas piping systems in the building. It also includes items such as domestic

---

1. NFPA is the National Fire Protection Association. This note and other fire suppression notes are available only with RSMeans Online, not the print book.

water softeners, swimming pool plumbing systems, drinking fountains and water coolers, and many other items. Plumbing work is indicated by fixtures on drawings, by a riser diagram, and by schedules. Many details are not shown, however, and are left to the estimator to include on the estimate, based on personal knowledge and experience. In many cases, water control, pipe, and fittings may be estimated as percentages of the fixture cost. In addition to the work shown on the drawings, the plumbing contractor may be required to make connections for building equipment.

When using RSMeans unit prices for plumbing fixtures, you have the option to look up each component of the connecting systems or search down the column from the fixture specified to use the first line you see marked "For rough-in, supply, waste, and vent." This line is predicated on the premise that most plumbing fixtures are mounted an average of 4 to 6 feet from a supply or waste line, which would be taken off separately. When lavatories and sinks are marked "with trim," they include a faucet set and drain tailpiece.

Fixture takeoff involves counting the various types, sizes, and styles. Be certain to include all costs, including trim, carrier, flush valves, and so forth, if the "rough-in, supply, waste, and vent" line is not used. Include equipment costs such as pumps, water heaters, water softeners, and all items not counted.

Piping is taken off by the system. Pipe runs for any type of system consist of straight sections and fittings of various shapes, styles, and purposes. The estimator will start at one end of each system and measure and record the straight lengths of each size of pipe. Fittings are noted and recorded as they are encountered. Care should be taken to note all changes in size and material. Such changes would occur only at a joint, so it should become a habit to see that the material of piping going into a joint is the same as that leaving the joint. It is good practice to round off the totals of each size of pipe to the lengths normally available from the supplier.

Nonspecialty estimators who are unfamiliar with the details of plumbing systems may choose to take off fixtures only and use general percentage markups for the various elements of the plumbing system. Percentages are suggested in R221113–40, Plumbing Approximations for Quick Estimating.

Heating, Ventilating, and Air Conditioning

The heating, ventilating, and air conditioning (HVAC) systems in the building are those systems that heat, cool, clean, humidify, or dehumidify the air. The actual systems used depend on the size and type of building, occupancy and use, the prevailing climate, and the type of equipment contained in the building. Mechanical work is usually performed by specialty subcontractors and estimated by specialists. The major subsections of Division 23, HVAC, are:

- Facility fuel systems, including piping and pumps
- HVAC piping and pumps
- HVAC air distribution
- HVAC air cleaning devices
- Central heating equipment
- Central cooling equipment
- Central HVAC equipment
- Decentralized HVAC equipment

As with plumbing, HVAC systems are shown primarily in riser diagrams and schedules. Systems are shown schematically, relying on the estimator's and installer's knowledge to fill in the standard details. Contract documents must be inspected with care and coordinated with the architectural, plumbing, and electrical drawings. Mechanical work requires substantial space within the building, and conflicts often occur.

HVAC equipment should be taken off first to familiarize the estimator with the building layout and the various mechanical systems. Items of equipment must be counted. Weight and size might be important if the unit is especially large or going into comparatively close quarters. If hoisting or rigging equipment is not owned by the mechanical contractor, then a rigging subcontractor will be required. From the equipment totals, other important items can be counted, such as motor starters, valves, strainers, gauges, thermometers, traps, air vents, and so forth.

HVAC sheet metal ductwork is usually estimated on a weight basis. This will involve taking off the lengths of the various sizes and converting to pounds per foot. A count must also be made of all the ductwork accessories, such as fire dampers, diffusers, registers, and so on.

The takeoff of heating and air conditioning pipe and fittings is done similar to plumbing. It is important to note whether joints and fittings are threaded, welded, or grooved. Many units are used in the takeoff within the HVAC system. Items such as boilers, chillers, fan-coil units, pumps, heat exchangers, and diffusers are taken off with a unit of each (Ea). Piping, round duct, and flex duct are taken off by LF. Metal ductwork is taken off by the pound (lb).

Example 1

Calculate the cost of a length of a rectangular galvanized steel duct 34″×20″×15′ long. Assume that there are approximately 2,500 lb of total weight of ductwork in the project. The duct is in a manufacturing facility located in Chicago, Illinois.

Solution

Ductwork is found in Section 23 31 13.13.

RSMeans Item No.	Description	Unit	Mat.	Labor	Equip.	Total	Total Incl. O&P
23 31 13.13 0570	Metal ductwork, fabricated rectangular, incl. fittings, joints, supports, and flexible connections, galvanized steel, 2,000 to 5,000 lb	lb	0.53	6.88	-	7.41	11.13

First, calculate the weight per LF of duct using R233100–40. Enter the chart at the top at the row, "SMACNA Max. Dimension—Long Side." Our long-side dimension is 34 inches. Because this is greater than 30 inches and less than 54 inches, the 54″ column should be used. Therefore, from the chart, the gauge of the sheet metal will be 22 ga. Next, enter the chart on the left-hand column, Sum-2 sides. Our duct is 34″×20″, so the sum of two sides is:

$$34″ + 20″ = 54″$$

On the 54″ row, the entry under the 22 ga. column is 13.5 lb per foot. The total weight of the 15-foot-long section is:

$$13.5\,lb/LF \times 15\,LF = 202.5\,lb$$

Note: *Table R233100–40 includes an allowance for hangers, slip joints, scrap, and so on. When the weight for each gauge of insulated duct is totaled, dividing the total by the weight in pounds per square foot for that gauge gives the total surface area in square feet. The quotient is the number of square feet of insulation required.*

The bare cost of the duct, from 23 31 13.13 0570 is:

$$202.5\,lb \times \$7.41 = \$1,500$$

The total cost with overhead and profit is:

$$202.5\,lb \times \$11.13 = \$2,254$$

The mechanical contractor usually is responsible for several subcontractors for items such as pipe insulation, duct, and equipment and controls systems. Having these subs may call for an additional set of markups for overhead and profit to be included in the estimate. When pricing the cost of HVAC systems, balancing air and water is included in the installation costs. If balancing by an independent contractor is specified in the contract, items in Sections 23 05 93.10 and 23 05 93.20 should be added to the estimate.

Electrical

The electrical system in a building consists of a service entrance, transformers and switchgear, power distribution, and lighting. The electrical work in the contract documents is shown by plans, schematic diagrams, and schedules. Many items may not be shown in detail, so the estimator must have knowledge of the components of each system and the methods of construction. The contract documents must include all components of the building electrical system that provide power to general building services, such as lighting, heating, ventilation, air conditioning, and general electrical wiring to outlets.

Many estimators begin the electrical takeoff by counting all the lighting fixtures, then adding columns to account for necessary components, including raceways, fixture whips, boxes, and switches. You can take advantage of any symmetry or repetition on different floors. Wire is usually figured by multiplying the measured conduit footages by the overall average of the number of wires contained in all the conduit runs. Since wire length must exceed conduit lengths in order to be pulled and to make connections and terminations, add 10 percent to the wire totals for waste and tie-ins. The measured conduit itself is usually increased by 5 percent to allow for installation adjustments and scrap.

Example 2

Calculate the cost of pulling three parallel copper wires, Type THHN, stranded #12 in ¾-inch electric metallic tubing (EMT) 240 feet long. The project is located in Bangor, Maine.

Solution

Wire quantities are taken off either by measuring each cable run or by multiplying the conduit and raceway quantities times the number of conductors in the raceway. Add 10 percent for waste and tie-ins. Keep in mind that the unit of measure of wire is usually 100 linear feet (CLF) for the smaller gauges (not linear feet [LF], as in raceways), so the formula would read:

$$\text{Wire quantity in CLF} = \frac{(\text{LF raceway} \times \text{No. of conductors}) \times 1.10}{100}$$

RSMeans Item No.	Description	Unit	Mat.	Labor	Equip.	Total	Total Incl. O&P
26 05 33.13 5020	Electric metallic tubing, ¾" diameter	LF	0.97	2.60	–	3.57	4.95
26 05 19.90 1200	Wire, copper, Type THWN-THHN, stranded, #12	CLF	10.37	30.63	–	41.00	57.30

To calculate the quantity of wire:

$$\text{Wire quantity} = \frac{(240\,\text{LF} \times 3) \times 1.10}{100}$$
$$= 7.92\,\text{CLF}$$

The bare cost of the wire is:

$$7.92\,\text{CLF} \times \$41.00 = \$324.72$$

The bare cost of the EMT is:

$$240\,\text{LF} \times \$3.57 = \$856.80$$
$$\text{Total bare cost} = \$324.72 + \$856.80 = \$1,182$$

The cost with overhead and profit is:

$$\$57.30 \times 7.92\,\text{CLF} + \$4.95 \times 240\,\text{LF} = \$1,642$$

Communications

Communications systems include data, voice, audio-video, structured cabling and enclosures, and wiring for voice and video communication, including telephone, television, public address systems, clock systems, and Internet wiring. Work is shown on plans by schematic diagrams and schedules. Special care should be taken to refer to the specifications for the detailed requirements of these systems and their integration into the building. Communications are usually installed by a specialty contractor, different from the electrical subcontractor. This might add an additional layer of overhead and profit to the costs.

Electronic Safety and Security

Electronic safety and security systems include access control, intrusion detection, video surveillance, fire detection and alarm, and fuel-gas detection. Work is shown in plans, schematic diagrams, and schedules. Special care should be taken to refer to the specifications for the detailed requirements of these systems and their integration into the building.

Chapter 11 Exercises—Set A

For solutions to the exercises in Set A, see the link to the student companion website at www.wiley.com/go/constructionestim5e. Some of these problems require using relevant reference notes in the RSMeans Building Construction Cost Data.

1. A plumbing contractor in Newark, New Jersey, needs to install 42 wall-hung, flush-valve-type toilets in a new office building.

   A. What is the cost of these fixtures and their rough-in?

   B. What would be the cost, including an allowance for water control, pipe, and fittings, if the building's plumbing is in a concentrated area?

2. A contractor in New Orleans, Louisiana, will be installing a gas-fired water heater in a new Cajun restaurant. It is expected that 250 to 300 meals will be served between 6 and 10 p.m.

   A. What size water heater will be required?

   B. How much should the contractor charge the GC for the installation?

3. A contractor needs to install a sprinkler system in an automobile repair shop. What system classification would you select?

4. A gymnasium is being built in Juneau, Alaska. The men's locker room will contain 12 showers, and the women's will have 8. There will also be 15 lavatories, 20 toilets, 4 urinals, 1 service sink, and 2 clothes washers in the facility. (Do not include plumbing cost for washers.) Usage will be steady throughout the day. What is the cost of installing the fixtures and an oil-fired water heater? Include an allowance for water control, pipe, and fittings.

5. A new apartment building in Washington, DC, has three levels of underground parking. The building footprint is 26,000 SF, and the floor-to-floor height is 9′–4″. How many cubic feet per minute (CFM) of air must be supplied? What would the appropriate fans cost the GC assuming the building to be in Milwaukee, Wisconsin? Use R233400–10. Garage use ranges 2 to 10 minutes. Use average time of 6 minutes/change.

6. If the fans are installed in May 2021, what will be the cost to the GC? Assume inflation at 3.5 percent.

7. How many BTUH (BTU/hour) and MBH (MBH = 1000 BTU/hour) are required to heat a 45,000-SF factory in Dover, Delaware? The building is two stories and has a flat roof. The floor-to-floor height is

16'–8". What would a gas/oil-fired hydronic boiler cost to heat this building? Use an outside design temperature correction factor for 10°F.

8. Calculate the cost of pulling four parallel copper wires, type THHN, stranded #10 in 1-inch EMT 600 feet long. The project is located in San Francisco, California. See R260519–90.

9. Calculate the cost of installing recessed 2'×4' fluorescent fixtures (with three 40-W lamps) in the office areas of a new office building in Denver, Colorado. Total office area is 30,000 SF, with a ceiling height of 10 feet. Use two fixtures per 100 SF. Refer to R265113–40.

10. Calculate the cost of a diesel generator for a building where the total electrical system is 2,000 kW and 20 percent of the system is on the emergency circuit. The project is located in Providence, Rhode Island.

Chapter 12

Earthwork
Division 31

Types of Earthwork

Earthwork includes many different kinds of work—from the initial site investigation to preparing the building or project for use. Along the way, a number of earthwork jobs must be done. These include:

- Subsurface investigation
- Excavation
- Tunneling
- Deep foundations (piles and caissons)
- Paving
- Piped utilities
- Sewerage and drainage
- Landscaping

In the CSI MasterFormat 1995, items such as demolition (including removing hazardous materials) and site preparation were included in Division 2 (Site Construction). In the 2004 MasterFormat (and its updates), demolition and site preparation were kept in the new Division 2 (Existing Conditions), while earthwork was moved to Division 31 (Earthwork).

Equipment Selection

There are probably more types of equipment used in earthwork (MasterFormat 2016 Division 31) than for all the other divisions combined. Earthwork equipment can be very expensive, and selection can be critical to production. Contractors are often faced with the question, "Which equipment is most cost efficient for the job?" or, "Should I use the equipment I already have, even though it is not the most efficient, or should I acquire (purchase, lease, or rent) the optimum equipment?"

With earthwork, the contractor's choice of equipment and the productivity of the crew depend on many factors, both job and business related. Job-related factors include type of materials (common earth, silt, clay, soft rock, hard rock, etc.), moisture content and denaturing requirements, weather, job site congestion, site access, size of the job, depth of excavation, length of haul for disposal, condition of the haul road, government regulations on disposal, and others. Business-related factors relate directly to the contractor's current workload and financial situation:

- Equipment availability
- Ability to finance new equipment
- Length of need period
- Accessibility of maintenance and repair service
- Availability and cost of labor

Refer to Chapter 13, "Equipment Analysis," for more on these issues.

Soil Excavation

Swelling and Shrinkage

When soil is excavated, it swells by a percentage related to the type of soil, degree of loosening, and moisture content. The volume of undisturbed earth (before excavation) is called *bank measure* (BM). The volume of the swelled soil after excavation is called *loose measure* (LM). If soil is compacted, its volume will shrink by a percentage depending on soil type and gradation and by moisture content, equipment used, depth of fill per lift (layer), and density required. (See Reference Number R312323–30.) The volume of compacted soil is called *compact measure* (CM). Since the cubic yard (CY) is the most commonly used unit in measuring the volume of soil, the three measuring units will be: BCY, LCY, and CCY, for bank, loose, and compact cubic yards, respectively (Figure 12.1). (S_w stands for swell factor and S_h for shrinkage factor.)

Bank volume	$= \text{Loose volume}/(1 + S_w/100)$
Bank volume	$= \text{Compact volume}/(1 - S_h/100)$
Loose volume	$= \text{Bank volume} \times (1 + S_w/100)$
Loose volume	$= \text{Compact volume} \times (1 + S_w/100)/(1 - S_h/100)$
Compact volume	$= \text{Bank volume} \times (1 - S_h/100)$
Compact volume	$= \text{Loose volume} \times (1 - S_h/100)/(1 + S_w/100)$

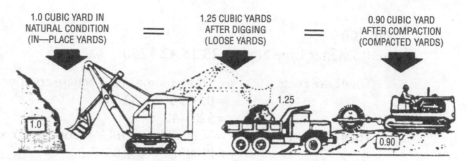

Figure 12.1
Reprinted from *RSMeans Heavy Construction Handbook*, by Richard C. Ringwald, PE. Used with permission.

Example 1

A contractor is to excavate an area of $200' \times 150'$ to a depth of $8'$ and haul it by dump truck to a site where it is to be used for fill. The soil will then be fully compacted. If the swell factor is 22 percent and the shrinkage factor is 15 percent, what is the total volume of the soil in BCY, LCY, and CCY?

Solution

Volume to be excavated	$= \text{Length} \times \text{Width} \times \text{Height}$
	$= 200' \times 150' \times 8'$
	$= 240{,}000 \text{ CF}$
Volume in bank measure	$= \text{Volume in CF}/\text{CF per CY}$
	$= 240{,}000/27$
	$= 8{,}889 \text{ BCY}$
Volume in loose volume	$= \text{Volume in bank measure} \times (1 + S_w/100)$
	$= 8{,}889 \times (1 + 22/100)$
	$= 10{,}844 \text{ LCY}$
Volume in compact volume	$= \text{Volume in bank measure} \times (1 - S_h/100)$
	$= 8{,}889 \times (1 - 15/100)$
	$= 7{,}556 \text{ CCY}$

Example 2

Estimate the total bare cost and total cost, including overhead and profit for the job in Example 1 (excavation only) using a 2½ CY front-end loader, track-mounted. Work will be in Erie, Pennsylvania.

Solution

Use RSMeans Line No. 31 23 16.42 1250:

Total bare cost	= Total volume in BCY × Bare cost per CY × CCI)
	= 8,889 BCY × \$2.30
	= \$20,445
Total cost incl. O & P	= Total volume in BCY × Cost per CY × CCI
	= 8,889 BCY × \$2.85
	= \$25,334

Example 3

What would the expected duration be for the job in Example 2?

Solution

Using line 31 23 16.42 1250, the daily output is 760 CY/day.

Duration in days	= Total quantity/Daily output
Duration	= 8,889 CY/760 CY/day = 11.7 ≈ 12 days

Example 4

The soil in Example 2 is to be hauled using 12 CY dump trucks for a distance of 2 miles (one way). What is the hauling bare cost and cost including overhead and profit (O&P)?

Solution

We have a large number of cost line items in *RSMeans Building Construction Cost Data* (*BCCD*) that are used for hauling with 12 CY dump trucks, starting from item 31 23 23.20 1014 to item 1714, depending on speed of truck and length of wait time (load, unload, and wait). Note that the distance mentioned represents the total cycle—that is, the round trip. For our example, let's assume average speed 25 miles per hour (mph) and 20 minutes wait time. Use item 31 23 23.20 1240.

Total bare cost	= \$1.96/CY labor + \$3.93/CY equipment = \$5.89/CY

Total bare cost = Total volume in LCY × Bare cost per CY
= 10,844 LCY × \$5.89 = \$63,871

Total cost incl. O&P = Total volume in LCY × Cost per CY
= 10,844 LCY × \$7.26/CY = \$78,727

Cost Basis For excavation, the cost is usually based on the bank measure (BCY). For hauling, it is based on LCY. For compaction, it is based on the CCY. Calculating the loose measure is important in case the soil has to be disposed of or imported and the number of truckloads must be calculated.

Example 5

A contractor is hired to excavate a trench 4′ wide, 5′ average depth, and 350′ long. An 18″-diameter pipe will be laid in, and then the trench will be backfilled and compacted (Figure 12.2). The swell factor is 22 percent and the shrinkage factor is 15 percent. Will the contractor need to borrow soil, dispose of excess soil, or neither?

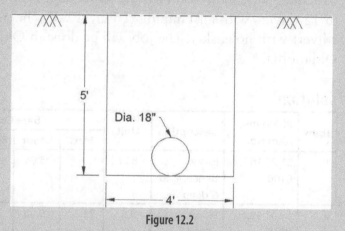

Figure 12.2

Solution

Volume of trench	$= \text{Length} \times \text{Width} \times \text{Height}$
	$= 350′ \times 4′ \times 5′$ $= 7,000\,\text{CF}$
Volume in BCY	$= \text{Volume in CF/CF per CY}$
	$= 7,000/27$ $= 259.3\,\text{BCY}$
Volume in LCY	$= \text{Volume in BCY} \times (1 + S_w/100)$
	$= 259.3\,\text{BCY} \times (1 + 22/100)$ $= 316.3\,\text{LCY}$
Volume in CCY	$= \text{Volume in BCY} \times (1 + S_h/100)$
	$= 259.3\,\text{BCY} \times (1 - 15/100)$ $= 220.4\,\text{CCY}$
Volume of pipe, CY	$= \text{Cross-section} \times \text{Length/CF per CY}$
	$= [\pi(18/12)^2/4] \times 350′/27$ $= 22.91\,\text{CY}$
Volume of compacted backfill	$= \text{Volume of trench} - \text{Volume of pipe}$
	$= 259.3 - 22.9$
	$= 236.4\,\text{CCY}$

As illustrated in the previous example, the excavated soil is insufficient, so the contractor will need to borrow.

Shortage (soil to be borrowed) = Excavated soil − Backfill volume
= 236.4 − 220.4
= 16 CCY

Shortage in loose measure = Shortage in CCY × (1 + S_w/100)/(1 − S_h/100)
= 16 × (1 + 22/100)/(1 − 15/100)
= 23 LCY

Example 6

Estimate the bare cost and the cost including O&P for performing the job in Example 5. Assume that excavation will be done with a ⅝ CY backhoe. Backfilling will be done by hand, and compaction will be in 12″ layers with air tamping (Figure 12.3). The pipe is a reinforced culvert with no gasket. The job will be done in Oklahoma City, Oklahoma.

Solution

Item	RSMeans Item No.	Description	Unit	Bare Costs				Total Incl. O&P
				Mat.	Labor	Equip.	Total	
1	31 23 16.13 0100	Excav, trench, 4′ to 6′ deep	BCY	–	3.04	2.06	5.10	6.89
2	33 41 13.60 2030	Pipe, concr, reinf, 18′	LF	23.27	9.60	2.32	35.19	42.67
3	31 23 23.13 0015	Backfill, hand, lt. Soil	LCY	–	22.50	–	22.50	34.50
4	31 23 23.13 1000	Comp., 12′ air tamp	ECY*	–	5.55	0.63	6.18	9.19
5	31 05 13.10 0200	Borrow, comm, 2 mile	CY	12.35	2.17	3.73	18.25	20.99

Item	Dimensions			Qty.	Unit
	Length	Width	Height		
1	350′	4′	5′	259.3	BCY
2	350′	–	–	350.0	LF
3	–	–	–	316.3	LCY
4	–	–	–	236.4	CCY
5	–	–	–	23.0	LCY

*RSMeans uses the designation ECY (embankment cubic yard), which is same as CCY.

Unit Estimate Summary By Subdivision, Oklahoma City, OK

Line Number	Long Description	Quantity	Unit Of Measure	Extended Total	Extended Total OP
31					
310513100200	Soils for earthwork, common borrow, spread with 200 H.P. dozer, includes load at pit and haul, 2 miles round trip, excludes compaction	23.00	C.Y.	$419.75	$482.77
31231613010	Excavating, trench or continuous footing, common earth, 5/8 C.Y. excavator, 4' to 6' deep, excludes sheeting or dewatering	259.30	B.C.Y.	$1,322.43	$1,786.58
312323130015	Backfill, light soil, by hand, no compaction	316.30	L.C.Y.	$7,116.75	$10,912.35
312323131000	Backfill and compact, by hand, 12" layers, compaction in layers, air tamp, add	236.40	E.C.Y.	$1,460.95	$2,172.52
33					
334113602030	Public storm utility drainage piping, reinforced concrete pipe (RCP), 18" diameter, 6' lengths, class 3, excludes excavation or backfill, gaskets	350.00	L.F.	$12,316.50	$14,934.50
			Totals	$22,636	$30,289

Answers may be rounded to $22,600 and $30,300.

When the amount of borrowed earth is as small as 23 CY, the cost per CY is likely to be higher. The cost of mobilization and demobilization was not included.

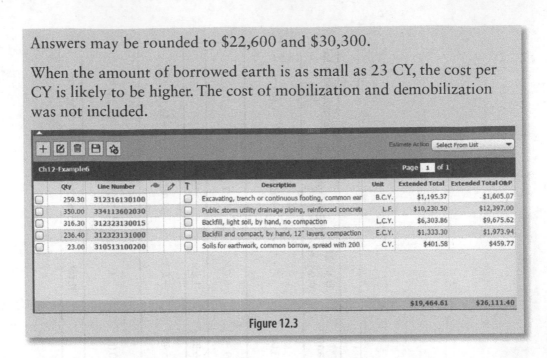

	Qty	Line Number			T	Description	Unit	Extended Total	Extended Total O&P
☐	259.30	312316130100			☐	Excavating, trench or continuous footing, common ear	B.C.Y.	$1,195.37	$1,605.07
☐	350.00	334113602030			☐	Public storm utility drainage piping, reinforced concrete	L.F.	$10,230.50	$12,397.00
☐	316.30	312323130015			☐	Backfill, light soil, by hand, no compaction	L.C.Y.	$6,303.86	$9,675.62
☐	236.40	312323131000			☐	Backfill and compact, by hand, 12" layers, compaction	E.C.Y.	$1,333.30	$1,973.94
☐	23.00	310513100200			☐	Soils for earthwork, common borrow, spread with 200	C.Y.	$401.58	$459.77
								$19,464.61	$26,111.40

Figure 12.3

Truck Capacity

Truck capacity, as used in the *BCCD*, represents the struck capacity, which is the volume the truck will hold when it is filled even with, but not above, the sides. Many contractors load their trucks to their heaped capacity, which is the volume the truck will hold when the earth is piled above the sides. Heaped capacity is greater than struck capacity by a margin that depends on the depth of the earth above the sides and the area of the truck bed. Truck capacity is specified in either volume (CY) or weight (tons).

Example 7

Rework Example 4 using heaped capacity. Assume the truck can take 20 percent more than its struck capacity. Also calculate the number of truckloads.

Solution

Truck heaped capacity	$= 12\ CY \times (1 + 20\%)$	$= 14.40\ CY$
Total soil to be hauled (from Example 4)		$= 10,844\ LCY$
Total bare cost	$= \$63,871/1.20$	$= \$53,226$
Total cost incl. O&P	$= \$78,727/1.20$	$= \$65,606$

The simplest way to calculate the cost would be to adjust the cost in proportion to the increase in capacity (e.g., 14.40 CY/12 CY = 1.20).

However, there is a small error in this assumption: The truck loading time will increase, causing the truck cycle time to slightly increase. This means the daily production's increase will be slightly less than the 20 percent increase in the truck capacity. Also, the truck's speed may be slightly affected. Let's put this argument in numbers:

The daily output for item 31 23 23.20 1240 = 192 CY/day using crew B34B with a daily cost of $1007.00 (bare) and $1256.46 (including O&P), before adjusting to location.

The truck original cycle time was (4 miles/25 mph) × 60 min/hour + 20 min wait time = 29.6 ≈ 30 minutes. So, in an 8-hour workday, there will be (12 CY × 8 hours/30 min) × 60 min/hour = 192 CY, which is same as the daily production shown for that item in RSMeans.

Now with heaped capacity, our earlier assumption is to add 20 percent to the daily production—that is, 230.4 CY. Assume the cycle time increased by 2 minutes to 32 minutes. The daily production = (14.4 CY × 8 hours / 32 min) × 60 min/hour = 216 CY, which represents only a 12.5 percent increase in daily production.

Based on this answer, we recalculate the cost (unit prices have been rounded to the nearest cent):

The CCI for Erie, Pennsylvania, Division 31, installation, is 103.2 percent and equipment is 117.3 percent.

Crew daily bare cost = $364.40 × 103.2% + $642.60 * 117.3% = $1,129.83
Crew daily cost incl. O&P = $549.60 × 103.2% + $706.86 * 117.3% = $1,396.33

Bare cost per CY	= Crew daily cost / Daily output	
	= $1,129.83/216 CY / day	= $5.23 / CY
Cost per CY incl. O&P	= Crew daily cost/Daily output	
	= $1,396.33/216 CY / day	= $6.46 / CY
Total bare cost	= 10,844 LCY × 5.23 / CY	= $56,722
Total incl. O&P	= 10,844 LCY × 6.46/CY	= $70,100
Number of truck loads	= Total quantity/Truck capacity	
	= 10,844 LCY/14.40 CY	
	= 753	

Duration of the job = 10,844/216 = 50.2 days. It is highly likely that this job will be done using several trucks[1] so the duration will be 50.2 divided by the number of trucks. With 4 trucks, the duration = 13 days.

1. See the section "Optimum Number of Trucks per Loader" later in this chapter.

Excavate by Hand or Machine?

Hand excavation is usually desirable over machine excavation in these cases:

- If the volume of excavation is too small to justify equipment mobilization and demobilization[2]
- In confined spaces where machines cannot reach or maneuver
- In areas where there are underground utilities (electric, gas, telephone, or other cables) that render machine excavation dangerous

Example 8

Hand excavation costs a contractor $40/CY, while machine excavation costs $14/CY. Transporting the excavation equipment to and from the job site costs about $600. For an excavation of a 12′ × 40′ area to a depth of 4 feet, should the contractor choose hand or machine excavation?

Solution

Total volume of excavation	= Length × Width × Depth	
	= 40′ × 12′ × 4′	= 1,920 CF
	= 1,920 CF/27 CF/CY	= 71.11 CY
Total cost with hand	= 71.11 CY × $40/CY	= $2,844
Total cost with machine	= 71.11 CY × $14/CY + $600	= $1,596

Machine excavation is more economical.

Example 9

A 4′-wide, 5′-average-depth trench has to be excavated. Hand excavation costs $32/CY. Machine excavation costs $6.50/CY in addition to $600 for transporting the excavation equipment to and from the site. What is the minimum trench length that justifies using the equipment?

Solution

Let L' be the length of the trench.

Total volume of excavation	= Length × Width × Depth	
	= L' × 5′ × 4′ CF	
	= L' × 5′ × 4′/27 CY	
	= 0.74L' CY	
Total cost with hand	= 0.74L' CY × $32/CY	= 23.68L'$
Total cost with machine	= 0.74L'CY × $6.50/CY + $600	= (4.81L'$ + 600)

---

2. When doing estimating in other countries, this point has to be studied carefully. Cheap labor and/or expensive equipment may justify larger hand excavation amounts.

Equating the two costs and solving for L':

$$\begin{aligned} 23.68L' &= 4.81L' + 600 \\ 18.87L' &= 600 \\ L' &= 31.8' \end{aligned}$$

Thus, for any length less than 32 feet, hand excavation will be more economical. Otherwise, machine excavation is the better choice. Other factors should be considered as well, particularly whether the equipment can be used for more work in the same project within the same time frame.

Example 10

Calculate the cost of excavating a trench, 4′ wide, 5′ deep, and 82′ long, in Los Angeles, California.

Solution

$$\begin{aligned} \text{Volume of excavation} &= \text{Length} \times \text{Width} \times \text{Depth} \\ &= 82' \times 5' \times 4'\ \text{CF} \\ &= 1{,}640\ \text{CF}/27\ \text{CF/CY} \\ &= 60.74 \approx 61\ \text{CY} \end{aligned}$$

Example 12-10 Page 1 of 1

Source	Qty	Line Number	Description	Freque	Crew	Unit	Labor Hc	Daily Outp	Extende	Extende	Extende	Extended Labor	Extended Labo	Extended Equip	Extended Equ	Extended Total	Extended Total O&P
Unit	61.00	312316130090	Excavating, trench or continuous footing, common earth		B11M	B.C.Y.	0.08	200.00				$246.44	$375.15	$122.00	$134.20	$368.44	$509.35
Unit	61.00	312316131400	Excavating, trench or continuous footing, common earth		1 Clab	B.C.Y.	1.00	8.00				$2,588.23	$3,982.08			$2,588.23	$3,982.08

Figure 12.4

$$\text{Unit bare cost} = \$42.43/\text{CY}$$

$$\text{Total bare cost} = 61\ \text{CY} \times \$42.43 = \$2{,}588$$

Next calculate machine excavation: Use line 31 23 16.13 0090 (Figure 12.4 also).

$$\text{Unit bare cost} = \$6.04/\text{CY}$$

Add mobilization: Use line 01 54 36.50 1300: Crew Equipment Cost Per Day = $258.45.

$$\text{Total bare cost} = 61\ \text{CY} \times \$6.04 + \$258.45 \times 2 = \$885$$

Note: The value for mobilization was doubled to include a charge for both mobilization and demobilization and because we assumed a 1-day cost per mobilization/demobilization.

In this example, many assumptions have been made. Some or all of the next points may be true:

1. Light soil

2. ½ CY backhoe

3. Mobilization within 50 miles

4. The cost of mobilization and demobilization of the backhoe was charged to this task only

Obviously, machine excavation is more economical in this case. However, every case has to be evaluated on its own merit.

Optimum Number of Trucks per Loader

In a situation requiring an excavator (such as a power shovel or a backhoe) and a number of dump trucks, the excavator will be working in almost the same spot, loading trucks. The trucks will travel to the dumping point, probably several miles away. Calculating the optimum number of trucks for the shovel is important. In the next example, we have assumed that cycle times for the loader and truck are constant.

Example 11

A ¾ CY power shovel is loading 12 CY dump trucks. (Assume that heaped capacity will make up for the swelling.) Trucks then haul the soil and dump it at a location 3 miles away. The average speed is 25 mph and the wait time is assumed to be 25 minutes.[3] Total quantity is 12,000 CY. Use national average prices.

Calculate:

1. Optimum number of trucks (number of trucks that cost minimum $/CY)

2. Total bare cost

3. Total cost including overhead and profit using RSMeans "Total Incl. O&P"

4. Total cost including overhead and profit using labor burden, plus 10 percent of all direct expenses for overhead and profit

---

3. According to RSMeans, "wait time" includes load, unload, and other wait time. In other words, the "wait time" includes everything except travel time.

Solution

Let's pick the next RSMeans items:

RSMeans Item No.	Description	Crew	CY/Day	Bare Costs				Total Incl. O&P
				Mat.	Labor	Equip.	Total	
31 23 16.42 3750	Excavating, bulk bank measure, ¾ CY, shovel	B12M	680	–	1.12	1.53	2.65	3.38
31 23 23.20 1442	Cycle hauling, excavated or borrow, LCY, 25 min wait/load/unload, 12 CY truck, cycle 6 miles, 25 mph	B34B	144	–	2.53	4.46	6.99	8.73

Productivity shown in the table for both items is based on the RSMeans assumption of a 50-minute work hour. We did not include any mobilization or demobilization cost.

For the loader (shovel), the crew used is:

Crew B12M:	Daily Bare Cost	Cost Incl. O&P
1 Equip. operator (crane)	$ 445.60	$ 674.00
1 Laborer	$ 313.20	$ 480.00
1 Crawler crane, 20 ton	$ 972.05	$ 1,069.26
1 FE attachments, 0.75 CY	$ 65.80	$ 72.38
Total	$1,796.65	$ 2,295.64

For the 12 CY dump trucks, the crew used is:

Crew B34B:	Daily Bare Cost	Cost Incl. O&P
1 Truck driver (heavy)	$ 364.40	$ 549.60
1 Dump truck, 12 CY	$ 642.60	$ 706.86
Total	$1,007.00	$1,256.46

Shovel cycle time:

Shovel's hourly production: 680 CY/day/8 hrs/day = 85 CY/hour

Loading Time   = Truck capacity/Loader output
= 12CY/85CY/hr × 60 min/hr = 8.47 minutes

Truck cycle time:

Position, load, unload, and wait (assumed)	= 25.0 min
Travel (haul and return):	
(6 mile × 60 min/hr)/(25 mile/hr)	= 14.4 min
Total	= 39.4 min

Number of trucks needed:

Truck cycle time/Shovel cycle time = 39.4/8.47 = 4.65 trucks

We have to use four or five trucks. From a theoretical point of view, it makes sense to use the number of trucks that gives the lowest cost per CY. From the practical point of view, however, it makes sense to use the higher number (e.g., five), because if one of the trucks breaks down for any reason, the fleet can still operate with close to its full production.

Daily production:

Since the shovel can load 4.65 trucks, when four trucks are used, the shovel will not operate at full capacity. It loads the four trucks in 8.47 × 4 = 33.88 minutes, then waits 39.40 – 33.88 = 5.52 minutes for the first truck to come back. The shovel's production in this case is 85 CY/hour × 4/4.65 = 73.12 CY/hr × 8 hrs/day = 585 CY/day. It can also be looked at from the truck's production point of view:

$$\text{Truck production} = \frac{12\,\text{CY/load} \times 60\,\text{min/hour}}{39.40\,\text{min/load}}$$
$$= 18.27\,\text{CY/hour}$$
$$\text{Four truck production} = 18.27\,\text{CY/truck/hr} \times 4\,\text{trucks}$$
$$= 73.1\,\text{CY/hr}$$
$$= 585\,\text{CY/day}$$

If we use five trucks, the first truck will come back and position itself while the shovel is still loading the fifth truck. The shovel will operate at full capacity, but the trucks will not. After the first load, each has to wait (5 × 8.47) – 39.4 = 2.95 minutes. The total fleet production will be the same as the shovel's—that is, 85 CY/hr = 680 CY/day.

Case 1: 4 Trucks
Total Daily Production = 585 CY
Daily costs:

Crew	Quantity	Bare Costs	Cost Incl. O&P
B12M (shovel crew)	1	$1,796.65	$2,295.64
B34B (truck crew)	4	$4,028.00	$5,025.84
Total		$5,825	$7,321

Bare cost per CY = **Crew daily bare cost/Crew daily output**
= $5,825/585 CY = $9.96

Cost per CY incl. O&P = $7,321/585 CY = $12.51

Case 2: 5 trucks
Total daily production = 680 CY
Daily costs:

Crew	Quantity	Bare Cost	Cost Incl. O&P
B12M (shovel crew)	1	$1,796.65	$2,295.64
B34B (truck crew)	5	$5,035.00	$6,282.30
Total		$6,832	$8,578
Bare cost per CY		= $6,832/680 CY = $10.05	
Cost per CY incl. O&P		= $8,333.5/680 CY = $12.61	

Using four trucks will result in slightly lower cost, but due to the very small difference, we recommend using five trucks. The next answers are based on a five-truck fleet.

Total bare cost = Total quantity CY × Bare cost per CY
= 12,000 CY × $9.96 = $119,520

Total cost including O&P = 12,000 CY × $12.51/CY = $150,120.

Equipment Rental Costs for Short or Long Periods

RSMeans equipment costs are based on the assumption of renting equipment for one week. For rental periods differing from one week, an adjustment has to be done to the equipment cost using equipment rental cost items in Section 01 54 33. For owned equipment, refer to Chapter 13 for hourly equipment cost, which consists of ownership and operation costs.

Example 12

A contractor in Los Angeles, California, needs to excavate 12,000 CY of heavy soil using a track-mounted front-end loader. He will rent the loader on a monthly basis. What is the total bare cost?

Solution

The loader in crew B10P, which is used in the items below, matches that in item 01 54 33.20 4530.

Using RSMeans lines 31 23 16.42 1300 and 31 23 16.42 4100, they can be combined as shown in Table 12.1 (note also the adjustment to the daily production):

Table 12.1

RSMeans Item No.	Description	Crew	CY/Day	Bare Costs				Total Incl. O&P
				Mat.	Labor	Equip.	Total	
31 23 16.42 1300	Excavating, bulk bank measure, 3 CY capacity, front-end loader, track mounted	B10P	1040	–	.61	1.25	1.86	2.30
31 23 16.42 4100	Excavating, bulk bank measure, for heavy soil or stiff clay, add						60%	60%
31 23 16.42 1300/4100	Excavating heavy soil, bulk bank measure, 3 CY capacity, front-end loader, track mounted	B10P	650	–	.98	2.00	2.98	3.68

From RSMeans item, 01 54 33.20 4530:

> Monthly rental cost = $11,548.20
> Hourly operation cost = $65.95

Assume 8 work hours per day and 21 workdays per month.

> Daily operating cost = $65.95/hr × 8 hrs/day = $527.60
> Daily total equipment cost = Daily operating cost
> + Monthly rental/Workdays per month
> = $527.60 + $11,548.20/21 = $1,077.51

> Daily production (adjusted for heavy soil) = 650 CY
> Equipment cost per CY = $1,077.51/650 CY = $1.66
> Labor cost per CY = $0.98
> Bare cost per CY = $2.64

Total bare cost in Los Angeles, California:

> Total bare cost = 12,000 CY × $2.64 = $31,680

Example 13

What is the total cost including O&P for the excavation in the previous example?

Solution

Since RSMeans unit costs including O&P combine material, labor, equipment, and O&P costs, we have to take them apart. RSMeans unit costs including O&P usually can be broken down as shown next:

Unit cost incl. O&P $= (M \times 1.1) + (L + \text{Column } F\%) + (E \times 1.1)$

where M, L, and E represent the bare unit cost for materials, labor, and equipment, respectively. "Column $F\%$" shows the total labor add-ons (taxes, insurance, overhead, and profit). It is part of the "Installing Contractor's Overhead & Profit" table on the inside of the back cover of the *BCCD* print book. In the Online Estimating, you can click on Reference, Labor Rates, Year 2017 Labor Rates, Standard Union Labor Rates (or other rates, depending on your selection). We also need the Workers'. Compensation Insurance reference note (R013113-60). The previous equation includes labor burden plus 10 percent of bare labor costs for profit. These costs are detailed in the "Installing Contractor's Overhead and Profit" table. Workers' compensation insurance rate for earth excavation in California is $9.0\% \times 2.33 = 20.97$ percent. We take the rest of labor add-ons from the table, columns C, D, and E. For type of labor, choose "Equipment Operator, Crane or Shovel" or "Equipment Operator, Medium."

Labor add-ons $= \text{Workers' comp} + \text{Average fixed overhead} + \text{Overhead} + \text{Profit}$
Labor add-ons $= 20.97\% + 18.3\% + 14\% + 10\% = 63.27\%$

Verifying unit cost, including O&P = ($0 × 1.1) + ($.98 × 1.6327) + ($2.00 × 1.1) = $3.80, which is slightly higher than the $3.68 mentioned in the table in Example 12. The fact that workers' compensation insurance rates in California are higher than the national average is probably one of the reasons.

Now let's use the numbers from Example 12 with the adjusted daily production:

Total cost incl. O&P in Los Angeles, California $= 12,000 \text{ CY} \times \$3.80 = \$45,600$

Example 14

Repeat Examples 12 and 13 based on daily rental.

Solution

Daily rental cost	= $1,291.58	
Daily operating cost	= $65.95/hr × 8 hrs/day	= $527.60
Daily total equipment cost	= Daily operating cost + Daily rental cost	
	= $1,291.58 + $527.60	= $1,819.18
Daily production (from last example)	= 650 CY	
Equipment cost per CY	= $1,819.18/650 CY	= $2.80
Labor cost per CY (same as last example)		= $0.98
Bare cost per CY	= $2.80 + $0.98	= $3.78
Total bare cost = 12,000 CY × $3.78		= $45,360
Total cost incl. O&P = 12,000 CY × ($.98 × 1.6327 + $2.80 × 1.1)		= $56,161

Note: The numbers in all of the preceding examples do not include transportation of the equipment to and from the site. This cost has to be taken into consideration.

Renting versus Owning Equipment

The decision whether to buy, lease, or rent equipment depends on several factors. The two most important issues are the extent of time the equipment is needed and the financial situation of the contractor. Generally speaking, if the contractor needs the equipment for a long time, buying is the best decision. Renting is the ideal solution for short-term needs. Leasing, if available, is an extended rental contract.

In the case of company-owned equipment, the RSMeans equipment cost has to be modified. The unit cost is calculated using these equations:

$$\text{Equipment cost per unit} = \frac{\text{Equipment daily cost}}{\text{Equipment daily output}}$$

$$\text{Equipment daily cost} = \frac{\text{Annual ownership cost}}{\text{Workdays per year}} + \text{Daily operating cost}$$

Annual ownership cost includes costs such as depreciation, maintenance, investment (interest), taxes, insurance, and storage. (For further discussion, see chapter 13 on equipment analysis.)

Chapter 12 Exercises—Set A

For solutions to the exercises in Set A, see the link to the student companion website at www.wiley.com/go/constructionestim5e.
Note: Answers do not include equipment mobilization and demobilization.

1. Estimate the total bare cost for building a 15,000-gallon septic tank in Huntsville, Alabama. Do not include excavation and piping.

2. Estimate the bare cost of excavation for the tank in Exercise 1, using a 1½ CY crawler-mounted backhoe. The tank is 15′ × 15′ cross-section and 10′ high. The excavation pit extends 2 feet from each side at the bottom, and the slope is 1:1. Depth of excavation is 14 feet. The soil is common earth. Hint: For the volume of excavation, use the formula:

$$\text{Volume} = \frac{1}{3} \times h \left(LB + UB + \sqrt{LB \times UB} \right)$$

where:

h = Vertical depth of the excavation
LB = Area of the lower base
UB = Area of the upper base

3. Estimate the total cost including O&P of both jobs in Exercises 1 and 2, using RSMeans Total Incl. O&P.

4. Estimate the total bare cost for drilling and blasting 2,500 CY of rock for a high-rise building foundation in downtown Memphis, Tennessee.

5. Estimate the total cost of Exercise 4 including O&P using RSMeans Total Incl. O&P.

6. What is the duration of the job described in Exercise 4?

7. What is the duration of the job if the crew worked 9 hours/day, 6 days/week?

8. What is the cost of performance bond for the job in Exercise 5?

9. A contractor contracted to excavate a trench 5′ wide, 5′ (average) deep, and 400′ long using a ¾ CY hydraulic backhoe. The contractor will lay down a 24″-diameter storm utility drainage pipe, reinforced concrete, no gaskets; then backfill and compact in 12″ layers using air tamping. The soil is common earth with a swell factor of 25 percent

and a shrink factor of 18 percent. The job will take place in St. Louis, Missouri. What is:

A. The amount of excess/shortage soil to be disposed/ borrowed?

B. The total bare cost for the operation?

C. The total cost including O&P?

10. A contractor has to excavate 7,500 CY of clay. He will be using 11 CY elevating scrapers and a pusher (dozer). The average haul distance is 1,500 feet. Work will be in Salt Lake City, Utah. What is:

A. The total bare cost?

B. The bare cost per CY?

C. The total cost and cost per CY including O&P using RSMeans Total Incl. O&P?

11. The contractor in Exercise 10 will employ a local open-shop work force. The local pay is $40/hour for an equipment operator and $32/hour for a laborer. What is the total bare cost?

12. What would the total bare cost and bare cost per CY in Exercise 10 be if the haul distance is 2,400 feet?

13. A 1½ CY power shovel is loading 12 CY dump trucks. Trucks then haul the soil a distance of 4 miles with an average speed of 25 mph and 20 minutes wait time. Swell factor is 20 percent, and trucks will be loaded at struck (12 CY) capacity. Total quantity is 22,000 CY. Work will take place in Tampa, Florida. Calculate:

A. Optimum number of trucks (i.e., number of trucks that gives you minimum $/CY)

B. Total direct cost

C. Total cost including overhead and profit

D. Duration of the job

14. Repeat Exercise 13, part B, based on these Tampa wages and assume an overhead and profit margin of 15 percent on all direct expenses:

Equipment operator: $39/hr
Equipment oiler or truck driver: $32/hr

15. Estimate the total bare cost for the job in Exercise 13 if the soil has to be graded by dozer at the dump site.

16. Estimate the total bare cost for excavating 8,000 CY of stiff clay in Topeka, Kansas, using a track-mounted front-end loader with a 2½ CY capacity.

17. Estimate the total bare cost and the total cost including O&P for excavating a utility trench, 12 inches wide, 24 inches deep, and 1.2 miles long, using a 40 H.P. chain trencher in Dallas, Texas.

18. Estimate the total bare cost and the total cost including O&P for preparing a crushed 1½-inch stone base, 6 inches deep, for a road segment 30 feet wide and 2 miles long in Phoenix, Arizona. Liquid asphalt emulsion will be applied at the rate of 0.3 gal/SY.

19. Estimate the total bare cost and the total cost including O&P for constructing 100 cast-in-place concrete friction piles, 50′ long, tapered steel, 4,000 psi, no reinforcing, 16″ diameter. Add mobilization costs. The job will take place in Chicago, Illinois.

20. Estimate the total bare cost and the total cost including O&P for constructing 50 treated wood piles, C.C.A. (chromate copper arsenate), 2.5# per CF, 10″ butts, 32′ long. Do not add mobilization costs. This job will take place in Buffalo, New York.

21. Estimate the total bare cost and the total cost including O&P for the excavation needed for the retaining wall in Exercise 1—Set A of Chapter 6. A ½ CY power shovel will be used. Assume that the excavation extends 3 feet from each end of the wall, with the same slope as the sides. Use the formula provided with Exercise 2 earlier.

22. A pit, 375′ long by 80′ wide by 2′ deep, will be filled and compacted in Seattle, Washington, with select structural fill. The swell factor is 20 percent, and the shrinkage factor is 15 percent. The soil will be borrowed from an area 5 miles away from the fill site. It will be compacted using sheepsfoot compactors. Estimate:

    A. The total volume of soil to be borrowed

    B. The total number of truckloads needed. Use 12 CY dump trucks, loaded with heaped capacity (110 percent of struck capacity)

    C. The total bare cost and total cost including O&P for the entire operation

23. Estimate the bare cost and the total cost including O&P for removing continuous concrete footing, 80′ long, 3′ wide, and 1′-6″ deep with average reinforcing in Toledo, Ohio. Disposal will be on site.

24. Estimate the bare cost and the total cost including O&P for supplying, filling, and compacting 5,500 CY of common fill in South Bend, Indiana. Soil will be hauled from a nearby quarry (about 1 mile away). Fill will be done by a dozer and compaction will be done with sheepsfoot roller. Shrinkage factor is 18 percent.

25. Estimate the bare cost and the total cost including O&P for asbestos removal in Parkersburg, West Virginia, that includes these items:

Type of Item	Quantity	Length, each
Beams W 18 × 40	20	35 LF
Beams W 30 × 108	12	42 LF
Duct insulation	1	1800 SF

Chapter 12 Exercises—Set B

Solutions to Set B exercises are available to instructors only; see the link to the instructor's companion website at www.wiley.com/go/constructionestim5e. Note: Total cost including O&P will be calculated using RSMeans Total Incl. O&P unless otherwise specified. Make any assumptions necessary for solving the problem.

1. Estimate the bare cost of the excavation for the retaining wall in Exercise 1—Set A of Chapter 6. Use the formula:

$$\text{Volume} = \frac{1}{3} \times h \left(LB + UB + \sqrt{LB \times UB} \right)$$

Where:

h = vertical depth of the excavation

LB = area of the lower base

UB = area of the upper base

Assume slope 1:1.

Use a 2-CY hydraulic backhoe, crawler-mounted. The location is Miami, Florida.

2. Estimate the total cost including O&P of the jobs in Exercise 1 using RSMeans Total Incl. O&P.

3. Estimate the total cost including O&P of the jobs in Exercise 1 using the answer of Exercise 1 plus all labor add-ons and 10 percent of direct expenses for overhead and profit.

4. Redo Exercise 1 assuming the excavator will be rented on a monthly basis and it will be used 20 days in the month.

5. Redo Exercise 1 with the productivity 15 percent better than that mentioned in the *BCCD*.

6. Redo Exercise 1 with workers working 10 hours per day and 5 days per week for the entire duration. Overtime hours are compensated at 1.5 times the regular pay. (You need to estimate the duration of the job in order to estimate the production efficiency.)

7. Redo Exercise 1 using the next local labor wages:

   A. $38/hour for equipment operator

   B. $32/hour for laborer

8. Redo Exercise 1 assuming the contractor is using a 1 CY hydraulic backhoe, crawler-mounted.

9. Estimate the bare cost for building a tennis playground with four courts and a total area of 120′ × 224′. The courts will be made of rubber-acrylic base resilient pavement. The project will be in Washington, DC.

10. Estimate the total cost—including O&P—of the job in Exercise 9 using RSMeans Total Incl. O&P.

11. Estimate the total cost—including O&P—of the job in Exercise 9 using the answer of Exercise 1 plus all labor add-ons and 10 percent of direct expenses for overhead and profit.

12. Redo Exercise 9 assuming the tandem roller will be rented on a daily basis.

13. Redo Exercise 9 with the productivity 15 percent less than that mentioned in the *BCCD* or RSMeans online estimating.

14. Redo Exercise 9 with workers working 11 hours per day and 5 days per week for the entire duration. Overtime hours are compensated at 1.5 times the regular pay. (You need to estimate the duration of the job in order to estimate the production efficiency.)

15. Redo Exercise 9 using these local labor wages:

Labor foreman:	$40/hour
Laborers:	$32/hour
Equipment operator:	$36/hour

16. Add a 10-foot-high chain-link fence for the courts in Exercise 9. The fence will apply only to the outside boundaries of the playground (120′ × 224′). Use 11-gauge wire, 2½″ post 10′ OC (on center). Add 8 gates, 4′ wide and 7′ high. Estimate the bare cost and total cost including O&P.

17. A contractor has to excavate 10,000 CY of common earth. He will be using 14 CY self-propelled scrapers and a pusher (dozer). The average haul distance is 3,000 feet. Work will be in Nashville, Tennessee, this coming summer. What is the total bare cost?

18. Estimate the total cost including O&P for the job in Exercise 17. Use RSMeans Total Incl. O&P.

19. Estimate the total cost including O&P for the job in Exercise 17 using the answer of Exercise 17 plus all labor add-ons and 10 percent of direct expenses for overhead and profit.

20. Redo Exercise 17, assuming the equipment will be rented on a monthly basis, and it will be used 21 days in the month.

21. Redo Exercise 17 with the productivity reduced by 20 percent.

22. Redo Exercise 17 with workers working 10 hours per day and 6 days per week for the entire duration. Overtime hours are compensated at 1.5 times the regular pay. (You need to estimate the duration of the job in order to estimate the production efficiency.)

23. Redo Exercise 17 using these local labor wages:

    A. $32/hour for equipment operator

    B. $25/hour for laborer

24. What would the bare cost in Exercise 10 be if the haul distance is 2400 feet?

25. A ¾ CY power shovel is loading 8 CY dump trucks. Assume that heaped capacity will make up for the swelling. (The heaped capacity of the truck is equivalent to 8 BCY.) Trucks then haul the soil a distance of 1 mile with an average speed of 20 mph. Assume a wait time of 20 minutes. Total quantity is 12,000 CY. Work will take place in Columbia, South Carolina. Calculate:

    A. The optimum number of trucks (i.e., number of trucks that gives you minimum $/CY)

    B. The total direct cost

    C. The total cost including overhead and profit

26. Redo Exercise 25 if the haul distance is 1.5 miles each way.

27. Redo Exercise 25 based on the following Columbia, South Carolina, open-shop wages:

Equipment operator:	$35/hour
Laborer:	$28/hour
Truck driver	$32/hour

Chapter 13
Equipment Analysis

Equipment cost is an essential part of any construction project budget. Equipment—ranging in size from a small air compressor to a huge tower crane—may be owned by the contractor, rented, or leased. In all cases, equipment costs must be accounted for in the contractor's cost estimating and budgeting. It is important to keep in mind that equipment incurs cost whenever it is under the contractor's control, even if it is idle.

Equipment Depreciation

Depreciation refers to a decrease in value. Most assets, such as equipment, structures, or furniture, depreciate because of deterioration and, sometimes, obsolescence. From an economic point of view, depreciation is calculated for tax purposes, and the amount of depreciation taken should result in the maximum allowable reduction of income tax to the business owner. For this reason, it is important to notice that the *book value*, which reflects the value of the equipment or asset in the accounting books based on depreciation calculations, may not reflect the true market value of that equipment or asset. Also, *useful life* according to accounting books might not be equal to the real useful life.

There are several methods of calculating depreciation, discussed in the next sections.

Straight-Line Depreciation

$$\text{Depreciation charge per year} = \frac{P - S}{n}$$

where:
P = purchase price including tax and shipping
S = salvage value at the end of useful life
n = useful life in years

If the equipment has tires, blades, drill bits, or other attachments that have a useful life different from the useful life of the equipment, these items must be depreciated separately.

Sum-of-Years-Digits Depreciation

Depreciation charge for any year =

$$\frac{\text{Remaining depreciable life at beginning of year}}{N} \times (P - S)$$

where:

$N = \Sigma$ years digits for total useful life

$N = 1 + 2 + 3 + \ldots + n = n[n+1] / 2$

P, S, n as defined earlier

Double Declining Balance Depreciation

$$\text{Depreciation charge per year} = \frac{2}{n} \times (P - \text{Depreciation charge to date})$$

$$= \frac{2}{n} \times (\text{Value at beginning of year})$$

When using the double declining balance (DDB) method, be careful not to go below the salvage value. Since this method does not consider the salvage value, there are, theoretically, three possibilities:

1. The depreciation amount lowers the value of the equipment to below the salvage value *before* the equipment reaches the end of its useful life.

2. The depreciation amount is not enough to get the value of the equipment down to the salvage value at the end of its useful life. (This is typically the case when salvage value = $0.)

3. In unusual cases, the depreciation amount will make the value of the equipment exactly equal to the salvage value at the end of its useful life.

When the depreciation amount is not enough to cause the equipment value to reach the salvage value at the end of the useful life, the DDB method may be used with a conversion to the straight-line (SL) method.

Double Declining Balance with Conversion to Straight-Line Depreciation

A comparison between DDB and SL depreciation is done on a year-by-year basis to determine which method provides the greatest tax advantage while being feasible. Once the SL depreciation becomes greater than the DDB depreciation, convert to the SL method. Example 2 explains this concept.

Modified Accumulated Cost Recovery System (MACRS) Depreciation

This is the newest depreciation method. In 1971, the US Treasury Department published guidelines for about 100 broad classifications of depreciable assets.

For each classification, there was a lower limit, midpoint, and upper limit of useful life, called the asset depreciation range, or ADR. In MACRS, property (equipment) is classified into classes. This method accelerates depreciation because the ADR midpoint lives are somewhat shorter than the actual average useful lives. Also, MACRS property classes are shorter than the ADR midpoint lives.[1]

Some methods are more accelerated than others. There is no "best" method for all cases. The contractor has to choose the method that works best for the situation. For example, if a contractor bought several new pieces of equipment in a given year, the net income in the first year after the purchase is probably small compared to total expenses. The contractor might not need to use an accelerated method to gain maximum tax benefit. The straight-line method might be most appropriate. On the other hand, if a contractor has large net income compared to expenses, it might be advantageous to take as much depreciation as possible. The next example uses several methods.

Example 1

A contractor purchased a bulldozer at a cost of $125,000. He has to pay $1,600 freight charge and 6 percent sales tax. The useful life for this bulldozer is 5 years, and the salvage value at that time is $20,000. What is the annual depreciation?

Solution

$$\text{Depreciation charge per year} = \frac{P-S}{n}$$

Cost f.o.b.[2]	$125,000
Freight	$ 1,600
Sales tax[3]	$ 7,596
Total without salvage	$134,196
Salvage value	−$20,000
Total with salvage	$114,196

---

1. For more details, see Donald Newnan, Ted Eschenbach, and Jerome Lavelle, *Engineering Economic Analysis*, 13th ed. (New York: Oxford University Press, 2017); and also IRS Publication 946.

2. Free on board (f.o.b.) could be either f.o.b. shipping point or f.o.b. destination. The term "f.o.b. shipping point" means delivered by the seller aboard the train, ship, etc., at the point of shipment, without charge to the buyer. However, the buyer has to pay for shipping from the point of shipment to the final destination. The term "f.o.b. destination" means delivered by the seller to the final destination, without charge to the buyer. In this book, the first interpretation will be used.

3. Some states tax shipping or freight charge and some do not.

A. Using straight-line depreciation:

Depreciation charge per year	=$114,196/5	=$22,839
The value at the end of the first year	=$134,196 – $22,839	=$111,357
The value at the end of the second year	=$111,357 – $22,839	=$88,518
The value at the end of the third year	=$88,518 – $22,839	=$65,679
The value at the end of the fourth year	=$65,679 – $22,839	=$42,840
The value at the end of the fifth year	=$42,840 – $22,839	=$20,000

B. Using sum-of-years-digits depreciation:

Depreciation charge for any year =

$$\frac{\text{Remaining depreciable life at beginning of year}}{N} \times (P - S)$$

$$n = 1 + 2 + 3 + 4 + 5 = 15 \text{ or } n = 5(5+1)/2 = 15$$

Depreciation charge for first year	=(5/15)×$114,196	=$38,065
Value at the end of first year	=$134,196 – $38,065	=$96,131
Depreciation charge for second year	=(4/15)×$114,196	=$30,452
Value at the end of second year	=$96,131 – $30,452	=$65,679
Depreciation charge for third year	=(3/15)×$114,196	=$22,839
Value at the end of third year	=$65,679 – $22,839	=$42,840
Depreciation charge for fourth year	=(2/15)×$114,196	=$15,226
Value at the end of fourth year	=$42,840 – $15,226	=$27,614
Depreciation charge for fifth year	=(1/15)×$114,196	=$7,614
Value at the end of fifth year	=$27,614 – $7,614	=$20,000

C. Using double declining balance depreciation:

Depreciation charge in any year

$$= \frac{2}{N} \times (P - \text{Depreciation charge to date})$$

Depreciation charge for first year	=($134,196)2/5	=$53,678
Value at the end of first year	=$134,196 – $53,678	=$80,518
Depreciation charge for second year	=($80,518)2/5	=$32,207
Value at the end of second year	=$80,518 – $32,207	=$48,311
Depreciation charge for third year	=($48,311)2/5	=$19,324
Value at the end of third year	=$48,311 – $19,324	=$28,987
Depreciation charge for fourth year	=($28,987)2/5	=$11,595

However, this depreciation puts the equipment value at $17,392, which is below salvage value. We take the fourth-year depreciation as $8,987 only.

Value at the end of fourth year = $28,987 – $8,987	= $20,000
Depreciation charge for fifth year	= $0
Value at the end of fifth year	= $20,000

Example 2

Assume that the equipment in Example 1 has no salvage value. Depreciate that equipment using the DDB depreciation method with conversion to the SL depreciation method.

Solution

Year	DDB depreciation*	SL depreciation*	Method	Value after depreciation
1	$134,196 × 2/5 = $53,678	$134,196/5 = $26,839	DDB	$134,196 − $53,678 = $80,518
2	$80,518 × 2/5 = $32,207	$80,518/4 = $20,130	DDB	$80,518 − $32,207 = $48,311
3	$48,311 × 2/5 = $19,324	$48,311/3 = $16,104	DDB	$48,311 − $19,324 = $28,987
4	$28,987 × 2/5 = $11,595	$28,987/2 = $14,494	SL	$28,987 − $14,494 = $14,493
5	$14,493 × 2/5 = $5,797	$14,493/1 = $14,493	SL	$14,493 − $14,493 = $0

* Values taken from solution of part C of Example 1.

Equipment Expenses

Equipment expenses are divided into two categories: ownership and operating expenses. The main difference between the two cost categories is that ownership cost is incurred whether the equipment is used or not. These costs include taxes, storage, interest, and insurance. Operating expenses, such as fuel, oil, grease, and tires, by contrast, are directly related to the use of the equipment. Maintenance cannot be clearly defined as ownership or operating cost, as it is both. Minimum maintenance is needed, even if the equipment is sitting idle. However, the more the equipment is used, the more maintenance costs. For rented equipment, in most cases, the owner pays ownership costs while the user (renter) pays operating expenses. For simplicity in this book, maintenance will be considered as an ownership cost.

Equipment expenses consist of:

- *Depreciation.* As explained earlier.
- *Maintenance.* Usually taken as 50 to 100 percent of the straight-line depreciation cost. Maintenance costs start low, then increase year after year as the equipment ages. However, spreading the maintenance cost uniformly over the useful life of the equipment—equivalent uniform annualized cost (EUAC)—is an acceptable practice for analysis.

- *Investment (interest)*. This expense is the "lost opportunity" cost—that is, the potential profit that would have been made if the money invested in equipment had been used in a different investment. Investment cost is calculated as the product of potential profit (or interest) multiplied by the average value of the equipment, P, which is calculated as:

$$\overline{P} = \frac{P(n+1) + S(n-1)}{2n}$$

  where:
  P = total initial cost
  S = salvage value at the end of useful life
  n = useful life in years

- *Taxes, insurance, and storage*. These expenses are taken either as a lump sum or as a percentage of the average value of the equipment.

- *Operating expenses, such as fuel, oil, other lubricants, tires, and other attachments*. Unless otherwise specified in the manufacturer's specifications, the average fuel consumption rate is 0.04 gallons per horsepower-hour for diesel engines and 0.06 for gasoline engines. To calculate the fuel consumption per hour, use the next equations:

For diesel engines:

$$\text{Fuel consumption} = 0.04 \times hp \times O_f \times T_f \times \text{Fuel cost per gallon}$$

For gasoline engines:

$$\text{Fuel consumption} = 0.06 \times hp \times O_f \times T_f \times \text{Fuel cost per gallon}$$

  where:
  hp = Rated force of the engine in horsepower
  O_f = Operating factor: the average power drawn as compared to the full power of the engine: usually about 50 to 60 percent
  T_f = Active time factor: actual work minutes per hour divided by 60

Oil hourly cost is taken as the total cost for an oil change divided by total hours between oil changes. The hourly depreciation cost of tires and other attachments is simply the cost of that item divided by its life in hours. Tire repair cost may be added as 10 to 20 percent of tire depreciation cost. When calculating the equipment depreciation, the initial cost of such items has to be subtracted from the total equipment cost. The cost of other lubricants, such as grease, is usually added as a percentage (around 50 percent) of oil hourly cost.

Example 3

Determine the probable cost per hour for owning and operating a crawler tractor for the listed conditions:

Engine	325 HP diesel
Shipping weight	69,900 lb
Freight rate	$2.75 per cwt (100 lb)
Capacity of crankcase	9 gallons
Estimated life	6 years
Estimated operating time	2,000 hours/year
Time between oil changes	120 hours
Time needed for oil change	4 hours
Shop cost of oil change	$75/hour
Assume a 55% operating factor and a 45-minute hour.	
Cost, f.o.b. factory	$300,000
Probable salvage value at the end of useful life	$32,000
Investment cost	14% of average value per year
Cost of diesel fuel	$2.75/gallon
Cost of lubricating oil	$10.00/gallon

Assume that maintenance and repairs cost 75% of straight-line depreciation.

Assume the annual cost for insurance, taxes, and storage is $7,500.

Solution

Cost, f.o.b. factory	$300,000
Freight: $699 \times \$2.75 =$	+ $1,922
Total	$301,922
Salvage value	− $32,000
Total	$269,922

$$\text{Average value } \bar{P} = \frac{(\$301,922 \times 7) + (\$32,000 \times 5)}{12} = \$189,455$$

Operating cost (hourly):

Fuel = 0.04 (factor for diesel engines) × 325 HP
 × 0.55 operating factor × (45 / 60) active time
 factor × $2.75 / gallon   = $14.75

Lubricating oil = (4 hours × $75 / hour + 9 gallons
 × $10.00 / gallon) / 120 hours between oil changes   = $3.25

Other lubricant = 0.5 × $3.25 (lubricating oil charge)   = $1.63

    Total operating cost   = $19.63 / hour

Ownership and other costs (annual):

Depreciation	= $269,922/6	= $ 44,987
Maintenance and repair	= $44,987 × 75%	= $ 33,740
Investment (interest)	= $189,455 × 14%	= $ 26,524
Insurance, taxes, and storage		= $ 7,500
	Total annual fixed cost	= $112,751
$Total fixed cost per hour	= $112,751/2000	= $56.38
Operating cost		+ $19.63
Total hourly cost		$76.01

Example 4

Determine the probable cost per hour for owning and operating a wheel-type tractor and scraper for the given conditions:

Engine	330 HP diesel
Shipping weight	64,000 lb
Freight rate	$2.50 per cwt (100 lb)
Capacity of crankcase	9 gallons
Estimated life	5 years
Estimated operating time	2,000 hours/year
Time between oil changes	100 hours
Time needed for oil change	4 hours
Shop cost for oil change	$75/hour
Assume a 60% operating factor and a 50-minute hour.	
Cost, f.o.b. factory	$295,000
Sales tax	6%
Cost of tires	$16,000
Life of tires	5,000 hours
Cost of repairing tires	15% of tire cost per hour
Probable salvage value at the end of useful life	$41,250
Investment cost	12% of average value per year
Cost of diesel fuel	$2.75/gallon
Cost of lubricating oil	$10.00/gallon

Assume that maintenance and repairs cost 80% of straight-line depreciation.

Assume the annual cost for insurance, taxes, and storage is $6,875.

Solution

Cost, f.o.b. factory		$295,000
Freight	$640 \times \$2.50$	$ 1,600
Sales tax	$\$295,000 \times 6\%$	+$ 17,700
Total		$314,300
Salvage value		– $ 41,250
Total without salvage		$273,050
Tires		– $ 16,000
Total without salvage and tires		$257,050

Operating cost (hourly):

Fuel	$330 \times 0.04 \times 0.60 \times (50/60) \times \$2.75$	= $18.15
Lubricating oil	$(4 \times \$75/hour + 9 \times \$10.00/gallon) / 100$	= $3.90
Other lubricant	$0.5 \times \$3.90$	= $1.95
Tire depreciation	$\$16,000 / 5,000$ hours	= $3.20
Tire repair	$\$3.20 \times 15\%$	= $0.48
	Total operating cost	= $27.68/hour

Ownership and other costs (annual):

Depreciation	= $ 257,050 / 5	= $ 51,410
Maintenance and repair	= $ 51,410 \times 80\%	= $ 41,128
Investment (interest)	= $ 205,080 \times 12\%	= $ 24,610
Insurance, taxes, and storage		= $ 6,875
Total fixed cost		= $124,023
Total fixed cost per hour	= $ 124,023 / 2,000	= $ 62.01
Operating cost per hour		= $ 27.68
	Total hourly cost	= $89.69

Example 5

How much would the cost of owning and operating the tractor and scraper in Example 4 be if the operating time is 1,500 hour/year?

Solution

Total fixed cost per hour	$ 124,023/1,500	$ 82.68
Operating cost per hour	(same as before)	$ 27.68
	Total hourly cost	$ 110.36

If the equipment is operated 1,500 hours per year, there will be less wear and tear than if it was operated 2,000 hours per year. It is expected that the maintenance and repair cost would be a little less than in the case of operating 2,000 hours per year, so a small adjustment (deduction) may be applied to the maintenance cost.

Equipment Rental

The decision whether to buy, lease, or rent equipment depends on several factors. The two most important issues are the extent of time the equipment will be needed and the financial situation of the contractor. Generally speaking, if the contractor needs the equipment for a long time, buying is best. Renting is the ideal solution for short-term needs. Leasing, if available, is an extended rental contract. For the options of buying or leasing, the contractor should try to keep the equipment as busy as possible to minimize the cost per hour. Many contractors rent or lend equipment that is not currently in use to fellow contractors who have temporary need for that equipment.

Rental rates depend on how long the equipment is needed. Basically, there are daily, weekly, and monthly rates. Monthly rates are the least costly per day, followed by weekly rates, then daily rates. When a contractor needs an equipment for three or four days, he should check on both the daily and weekly rates. He may be better off renting the equipment for a week, even though he needs it for only three days. Weekly rental rates in RSMeans Cost Works and *Building Construction Cost Data* (*BCCD*) are roughly equal to three times the daily rental rates. The same cost issue arises when the equipment is needed for 15 days; a comparison between weekly and monthly rates is necessary.

Rental rates in RSMeans online estimating and the *BCCD* apply to late-model, high-quality machines in excellent working condition, rented from equipment dealers. Rental rates from contractors may be substantially lower than rates from equipment dealers, depending on economic conditions. Rental rates for older, less productive machines may be lower by up to 15 percent.

In most rental agreements, the renter pays for the routine operating expenses, such as fuel, grease, and oil changes, if needed. The equipment owner pays for major maintenance, taxes, and insurance.

Subdivision 01 54 33 in the *BCCD* deals with equipment rental. There are five *cost* columns. The first column shows the hourly operating costs including fuel, oil, and routine maintenance. The next three columns display the daily, weekly, and monthly rental costs, respectively. The last column displays the "Crew Equipment Cost/Day." This represents the total *daily* cost, both operating and rental, based on a weekly rental and assuming five 8-hour days per week. For example, see item 01 54 33.20 0150, an excavator, 1 CY capacity, diesel hydraulic, crawler mounted: (Figure 13.1).

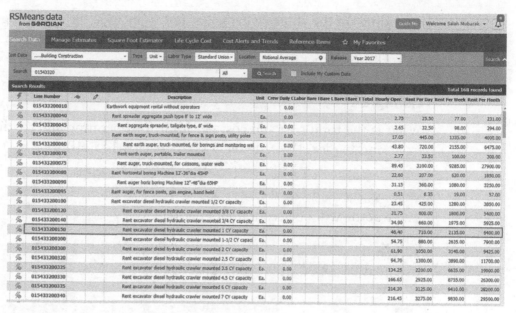

Figure 13.1

Daily rental cost based on weekly rental and:

Cost of rent per day based on weekly rental and 5 workdays per week

	= $2,135/5	= $427
Daily operating cost	= $46.40×8 hours/day	= $371.20
Crew equipment cost	= $798.20	

However, RSMeans Online estimating has a nice feature: When an equipment item is chosen to be included in the estimate, a dialog screen appears asking the user to fill in the following (Figure 13.2):

- Rental period: daily, weekly, or monthly
- Include hourly operating cost

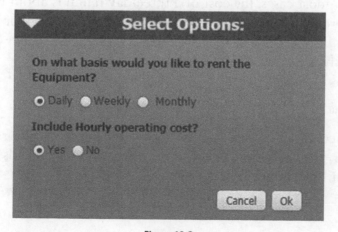

Figure 13.2

Depending on the choice whether or not to include the operating cost, the item description in the estimate will include "Excl. Hourly Oper. Cost" or "Incl. Hourly Oper. Cost."

Example 6

What is the total cost and cost per hour for operating and renting a front-end loader, 4WD, diesel, 3 CY capacity, 145 HP, for:

a. 2 days	b. 3 days	c. 4 days
d. 13 days	e. 1 month	

The tractor will be used 8 hours per day.

Solution

Use item 01 54 33.20 4710 (Figure 13.3).

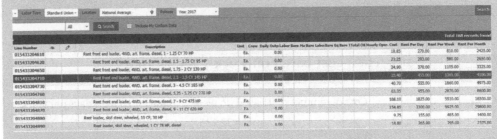

Figure 13.3

Operating cost per day		= $35.40/hour × 8 hours/day	= $283.20
Rental daily cost based on daily rental			= $455

Two days:

2 days' total cost	2 × ($283.20 + $455)	= $1,476.40
Total cost per hour	$1,476.40/16 hours	= $92.28

Three days: In most cases, the cost of a three-day rental is about the same as the weekly rental. In our case, the weekly rental is $1,365. For the sake of comparison, we'll use daily rent for three days (e.g., 3 × $455 = $1,365), which is exactly the same, so let's assume one-week rental.

Total cost	= $1,365 + (3 × $283.20)	= $2,214.60
Total cost per hour	= $2,067 / (3 days × 8 hours)	= $92.28

From a practical point of view, it is better for the contractor to rent the equipment for one week rather than three days, in case he needs to keep the equipment for another day.

Four days: Obviously, it will be better to rent the equipment for one week, even though it is needed for only four days.

Weekly rental cost		= $1,365
Total cost	= $1,365 + (4 × $283.20)	= $2,497.80
Total cost per hour	= $2,497.80 / (4 days × 8 hours)	= $78.06

Thirteen (working) days: Thirteen working days is almost equivalent to three weeks. Let's compare:

13 days' rent	= 13 × $455	= $5,915
3 weeks' rent	= 3 × $1,365	= $4,095
1 month's rent		= $4,100

For the same previous argument, it is better to rent for one month.

Total cost	= $4,100 + (13 × $283.20)	= $7,781.60
Total cost per hour	= $7,257 / (13 days × 8 hours)	= $74.82

One month: Assume 21 working days/month:

1 month's rent		= $4,100
Total cost	= $4,100 + (21 × $283.20)	= $10,047
Total cost per hour	= $10,047/(21 days × 8 hours)	= $59.80

Note: Rental rates in this example represent national averages. RSMeans Online estimating and the BCCD contain national average pricing, which can be adjusted for geographic location using city cost indexes or location factors.

Chapter 13 Exercises—Set A

For solutions to the exercises in Set A, see the link to the student companion website at www.wiley.com/go/constructionestim5e.

1. Determine the probable cost per hour for owning and operating a crawler tractor for the given conditions (use straight-line depreciation method):

Engine	250 HP diesel
Shipping weight	49,000 lb
Freight rate	$2.65 per cwt
Capacity of crankcase	7.5 gallons
Estimated life	5 year
Estimated operating time	1,800 hours/year
Time between oil changes	100 hours
Time needed for oil change	4 hours
Cost of labor for oil change	$70/hour
Assume a 60% operating factor and a 45-minute hour.	
Cost, f.o.b. factory	$188,000
Georgia sales tax	5% (on purchase cost only)
Probable salvage value	$0
Investment cost	10% of average value/year
Cost of diesel fuel	$2.50/gallon
Cost of lubricating oil	$8.50/gallon
Assume that repairs cost 75 percent of depreciation.	
Assume the annual cost for insurance, taxes, and storage is $6,000.	

2. Repeat Exercise 1 if the equipment has a $24,000 salvage value and maintenance cost is 100% of depreciation cost.

3. Repeat Exercise 1 (no salvage value) if the equipment is wheel-mounted and:

   A. Cost of tires: $12,000

   B. Life of tires: 5,000 hours

   C. Cost of repairing tires: 15% of tire cost per hour

4. Repeat Exercise 3 if the equipment has a salvage value of $20,000 and maintenance cost is 90% of depreciation cost.

5. Repeat Exercise 1 if the equipment will be used 1,200 hours/year only.

6. Depreciate the equipment in Exercise 1 using:

   A. Sum-of-years-digits method.

   B. Double declining balance method.

   C. Double declining balance with conversion to straight-line depreciation.

7. Determine the probable cost per hour for owning and operating a crawler tractor for the given conditions:

Engine	200 HP diesel
Shipping weight	40,000 lb
Freight rate	$2.50 per cwt
Capacity of crankcase	6 gal
Estimated life	6 yr
Estimated operating time	1,600 hr/yr
Time between oil changes	100 hours
Time needed for oil change	3 hours
Cost of labor for oil change	$80/hr

Assume a 60 percent operating factor and a 50-minute hour.

Cost, f.o.b. factory	$140,000

Minnesota sales tax 6.5% (on purchase and shipping cost)

Probable salvage value	$0
Investment cost	12% of average value per yr
Cost of diesel fuel	$2.60/gal
Cost of lubricating oil	$9/gal

Assume that repairs cost 80 percent of depreciation.

Assume the annual cost for insurance, taxes, and storage is $5,400.

Use straight-line depreciation method.

8. Repeat Exercise 7 if the equipment has a $24,000 salvage value and maintenance cost is 100 percent of depreciation cost.

9. Repeat Exercise 7 (no salvage value) if the equipment is wheel-mounted and:

   A. Cost of tires: $12,000

   B. Life of tires: 5,000 hours

   C. Cost of repairing tires: 15 percent of tire cost per hour

10. Repeat Exercise 9 if the equipment has a salvage value of $20,000 and maintenance cost is 90 percent of depreciation cost.

11. Repeat Exercise 7 if the equipment will be used 1,200 hours/year only.

12. Depreciate the equipment in Exercise 7 with a $20,000 salvage value and using:

   A. Double declining balance method.

   B. Sum-of-years-digits method.

13. Using US national average rates, calculate the total *hourly* cost for renting and operating a 100 KW generator, diesel engine. Make calculations based on:

   A. Monthly rental and 21 workdays per month.

   B. Weekly rental and 5 workdays per week.

   C. One-day rental.

   D. Four-day rental only.

   E. Six-hour rental only.

14. Using US national average rates, calculate the hourly cost to rent and operate (do not include operator) a crane, truck-mounted, cable-operated, 8 × 4, 15′ radius, 90-ton capacity, for a rental period of:

   A. 14 days.

   B. 3 days.

   C. 1 day only.

   D. 1 week with 5 regular workdays.

   E. 1 week with 10 hours/day and 6 workdays/week Assume an additional pay of $50/hour for overtime hours.

Chapter 13 Exercises—Set B

Solutions to Set B exercises are available to instructors only; see the link to the instructor's companion website at www.wiley.com/go/ constructionestim5e.

1. Determine the probable cost per hour for owning and operating a crawler tractor for the given conditions (use the straight-line depreciation method):

Engine	225 HP diesel
Shipping weight	45,000 lb
Freight rate	$2.95 per cwt
Capacity of crankcase	7 gal
Estimated life	5 yr
Estimated operating time	1,700 hr/yr
Time between oil changes	110 hours
Time needed for oil change	4 hours
Cost of labor for oil change	$90/hr
Assume a 55% operating factor and a 50-minute hour	

Cost, f.o.b. factory	$225,000
Florida sales tax 6%	(on purchase cost only)
Probable salvage value	$0
Investment cost	10% of average value per year
Cost of diesel fuel	$2.50/gal
Cost of lubricating oil	$12.00/gal
Assume that repairs cost 75% of depreciation	
Assume the annual cost for insurance, taxes, and storage is $6,800.	

2. Repeat Exercise 1 if the equipment has a $20,000 salvage value and maintenance cost is 90% of depreciation cost.

3. Repeat Exercise 1 (no salvage value) if the equipment is wheel-mounted and:

   A. Cost of tires: $10,000

   B. Life of tires: 4,500 hours

   C. Cost of repairing tires: 20% of tire cost per hour

4. Repeat Exercise 3 if the equipment has a salvage value of $20,000 and maintenance cost is 90 percent of depreciation cost.

5. Repeat Exercise 1 if the equipment will be used 1,500 hours/year only.

6. Depreciate the equipment in Exercise 1 using:

   A. Double declining balance method.

   B. Sum-of-years-digits method.

   C. DDB depreciation method with conversion to SL depreciation method.

7. Determine the probable cost per hour for owning and operating a crawler-mounted hydraulic backhoe with diesel engine for the given conditions (use the straight-line depreciation method):

Engine	300 HP diesel
Shipping weight	52,000 lb
Freight rate	$2.75 per cwt
Capacity of crankcase	8 gal
Estimated life	6 yr
Estimated operating time	1,500 hr/yr
Time between oil changes	120 hours
Time needed for oil change	5 hours
Cost of labor for oil change	$70/hr
Assume a 55% operating factor and a 50-minute hour	
Cost, f.o.b. factory	$200,000
Nevada sales tax	6.75% (on purchase cost only)

Probable salvage value	$0
Investment cost	11% of average value per year
Cost of diesel fuel	$2.50/gal
Cost of lubricating oil	$9.00/gal

Assume that repairs cost 75 percent of depreciation.

Assume the annual cost for insurance, taxes, and storage is $6,500.

8. Repeat Exercise 7 if the equipment has a $30,000 salvage value and maintenance cost is 90 percent of depreciation cost.

9. Repeat Exercise 7 (no salvage value) if the equipment is wheel-mounted and:

   A. Cost of tires: $15,000

   B. Life of tires: 6,000 hours

   C. Cost of repairing tires: 15 percent of tire cost per hour

10. Repeat Exercise 9 if the equipment has a salvage value of $25,000 and maintenance cost is 90 percent of depreciation cost.

11. Repeat Exercise 7 if the equipment will be used 1,200 hours/year only.

12. Depreciate the equipment in Exercise 7 with a $30,000 salvage value and using:

   A. Double declining balance method.

   B. Sum-of-years-digits method.

13. Calculate the total hourly cost for renting and operating a 240 HP sheepsfoot vibratory roller for a period of:

   A. Monthly rental and 20 workdays per month.

   B. Weekly rental and 5 workdays per week.

   C. 1 day only.

   D. 4 days only.

   E. 16 workdays.

14. Calculate the hourly cost to rent and operate (do not include operator) a chain trencher, diesel engine, 8′ deep, 16″ wide, for a rental period of:

   A. 14 days.

   B. 3 days.

   C. 1 day only.

   D. 1 week with 5 regular workdays.

   E. 1 week, with 10 hours/day and 6 workdays/week.

Assume an additional rental cost of $250/hour for overtime hours.

Chapter 14

Assemblies Estimating

In the early stages of project design, building professionals can use assemblies estimating (sometimes known as *systems estimating*) to develop a project budget and evaluate alternative building systems. *Assemblies estimates* provide quick, comprehensive ways to study the cost impact of building systems as they relate to the project budget.

Preliminary Cost Estimating

The assemblies approach to preliminary cost estimating involves grouping several trades and/or work items and tasks into a building element. An *assembly* is a grouping of unit price line items, with appropriate quantities, to form a building component, such as an exterior wall, partition, roof, or footing system. For example, a foundation wall usually requires formwork, reinforcing, concrete placement, and finish. A unit price estimate of the foundation wall would require that you evaluate and compile several separate line items for each of these activities to arrive at a cost. In an assemblies estimate, the separate line items are combined in a single package. The assembly cost of a foundation wall, then, is represented as one line item that includes the cost of materials and installation (labor plus equipment) applied to one unit/quantity.

Assemblies estimates do not require a completed design or detailed drawings. Instead, they are based on the general size of the structure and other known parameters of a project. The degree of accuracy of an assemblies estimate is generally plus or minus 10 percent.

Assemblies Estimates

Assemblies estimates are organized differently than the unit price estimates discussed earlier in this book. Because these estimates are performed during the schematic and design development phases of the project, the materials-based MasterFormat system does not work well. The UniFormat II system, based on major building elements such as walls, flooring, or HVAC systems, is more appropriate for assemblies estimating. In

UniFormat II, data are organized by where the item occurs in the building, in the general order of construction. Hence, Division A, Substructure, contains information on foundations. Division B, Superstructure, and covers the framing, exterior closure, and roofing. Division C covers interior finish, Division D covers services (conveying systems, mechanical, and electrical), Division E covers equipment and furnishings, Division F covers special construction, and Division G covers building sitework.

When using the RSMeans Online Estimating, you need to go to the Cost Data box and choose Assemblies, as shown in Figure 14.1.

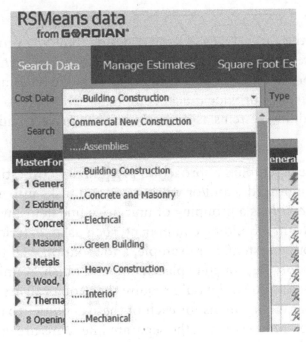

Figure 14.1

Example 1

Using RSMeans assemblies, calculate the cost of a foundation wall in a building in Boston, Massachusetts. Assume the wall to be 10 inches thick and 8 feet high. The concrete will be placed by direct chute. The wall rests on a strip footing that is 24 inches wide by 12 inches deep. Soil-bearing capacity is 6 KSF. The building perimeter is 200 LF.

Solution

Using RSMeans Online Estimating, first set the location to Boston, Massachusetts. Then select the Assemblies to access the assemblies cost data. Use the locator tree or the Search box to select the section for strip footings. Select assembly A 1010 110 2700 (Figure 14.2).

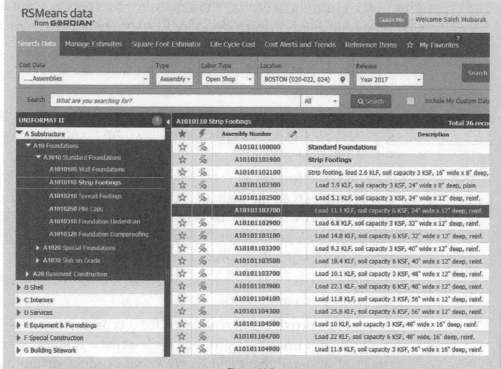

Figure 14.2

Add the item to the estimate and enter the quantity 200 in the quantity box. The extended costs for materials, installation, and total will appear. Repeat the procedure for the foundation wall, assembly number A 2020 110 5040 (Figure 14.3).

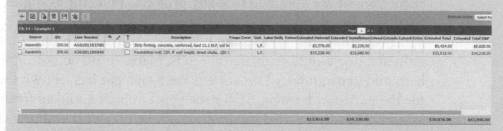

Figure 14.3

The total bare cost for our foundation wall on a strip footing in Boston, Massachusetts, is $30,036 ≈ $30,000. The total cost, including overhead and profit, is $43,046 ≈ $43,000. These numbers reflect open shop labor. If we use union labor, the cost increase to $36,992 ≈ $37,000 and $50,972 ≈ $51,000.

If you use the RSMeans print book *Assemblies Cost Data,* see "Using the Assemblies." The process of locating the correct assemblies, multiplying total costs by quantities, and then adjusting for location using the Division A city cost index will be similar with possible small difference in the cost numbers.

This quick example illustrates the key differences between the unit cost data found in *RSMeans Building Construction Cost Data* and the assemblies cost data:

- Assemblies represent building elements made up of several unit price items/tasks.

- Most assemblies are accompanied by an illustration, a written description, and a list of the assembly components with their quantities and costs.

- Assemblies information is far less specific than unit cost data. It is important to read through all the choices available in a given section and select the assembly that is closest to the intended work. Rarely will there be a perfect match. Select the assembly closest to the design intent, or interpolate between two similar assemblies.

- Assemblies costs are shown for material and installation (the sum of the labor and equipment).

- Assemblies provide no breakdown for a typical crew. (Assemblies are multiple unit price lines grouped to describe a building element. For the level of detail that includes crew and productivity, unit price data should be used instead.)

- Assemblies costs represent the total costs, including the overhead and profit for the installing contractor.

Assemblies cost data can also be used to speed up and simplify the work in a unit cost estimate. Assemblies are particularly useful for building elements that are common to many projects. Take, for example, an interior partition composed of 2×4 wood studs 16″ on center (OC), 3½″ acoustic batt insulation, and ½″ gypsum board on each side. To do a unit price estimate, one would estimate the studs, plates (top and bottom), acoustic batt, and gypsum board separately. Each element has its own unit: LF for the partition framing, SF for the gypsum board, and so forth. In the assemblies estimate, this partition would be taken off and priced as one element and measured with one unit, SF. Using the assembly wherever this common partition occurs in a unit price estimate can save a great deal of time without a significant loss of accuracy.

Combining Assemblies and Unit Costs

If you, as an estimator, are combining unit costs and assemblies costs in the same estimate, be careful to place all assembly cost data in the Total with Overhead and Profit column. If you are using RSMeans Online estimating, you must create two different estimates—one for items and one for assemblies—but the location must be the same. After saving them, go to Manage Estimates, choose both estimates (by checking the box on the far left next to each estimate) and choose the Combine function. This results in the creation on a new (third) estimate. You can do this also by exporting the files to a spreadsheet and combines them manually. Again, when combining the data on a single spreadsheet, be careful to put the costs of the assemblies in the Total with Overhead and Profit column.

Chapter 14 Exercises—Set A

For solutions to the exercises in Set A, see the link to the student companion website at www.wiley.com/go/constructionestim5e.

1. Estimate the cost of the next spread footings. The soil capacity is 6 KSF. The project will be in New York City, New York.

    A. 12 spread footings with a loading capacity of 500 kips each

    B. 20 spread footings with a loading capacity of 300 kips each

    C. 8 spread footings with a loading capacity of 150 kips each

2. Estimate the cost of a reinforced slab on grade, 120′ long, 56′ wide, 6″ thick, nonindustrial, in Chicago, Illinois.

3. Estimate the cost of a CIP waffle slab, 20′ × 25′ bays, 40 PSF superimposed load. Total area is 100′ × 50′. The building will be located in Phoenix, Arizona.

4. Estimate the cost of a standard exterior CMU block, 8″ × 16″ × 8″ with two coats cement stucco, ⅝″ thick. The wall is 8′ high and 80′ long, located in Miami, Florida.

5. Add the following items to the estimate in Exercise:

    • Precast lintel, 4′ long, 4″ wide, 8″ high, prestressed, 4 each

    • Precast lintel, 4′ long, 6″ wide, 8″ high, prestressed, 4 each

    Remember that you need to create two separate estimates (one for items and one for assemblies) and then merge them.

6. What is the cost of providing and installing 10 steel doors, 18 gauge hollow metal, single door, with frame, no label, 3′ × 7′ opening? The building is located in Nashville, Tennessee.

7. Estimate the cost of building and finishing interior partitions with a total of 1,200 SF for a building in Boise, Idaho, with the following specs:

    • ⅝″ fire-rated gypsum board on 2 × 4 wood studs spaced at 16″ OC. Same opposite face, no insulation

    • Painting, brushwork, primer, and two coats

Chapter 14 Exercises—Set B

Solutions to Set B exercises are available to instructors only; see the link to the instructor's companion website at www.wiley.com/go/ constructionestim5e.

1. Estimate the cost of the next spread footings. The soil capacity is 3 KSF. The project will be in Baton Rouge, Louisiana.

   A. 24 spread footings with a loading capacity of 200 kips each

   B. 16 spread footings with a loading capacity of 300 kips each

   C. 8 spread footings with a loading capacity of 500 kips each

2. Estimate the cost of a reinforced concrete strip footing, 75′ long, 16″ deep, and 48″ wide. The load is 22 KLF and the soil-bearing capacity is 6 KSF. The project will be in Tucson, Arizona.

3. Estimate the cost of a CIP flat plate (slab), 20′ × 20′ bays, 125 PSF superimposed load. Total area is 120′ × 80′. The building will be located in Fort Wayne, Indiana.

4. What is the estimated cost for four 50′-long steel pile clusters to carry 400 K load, each? The project is in St. Petersburg, Florida. Assume the piles are:

   A. End-bearing type (use 7 pile-clusters)

   B. Friction type (use 16 pile-clusters)

5. Estimate the cost of a concrete block (CMU) partition, regular weight, hollow, 8″ thick, with gypsum plaster. It will get two coats on the inside only. This will be the exterior for a warehouse building, 64′ long, 28′ wide, and 12′ high, and will be located in Worcester, Massachusetts. The building will have these openings:

Type	Quantity	Dimensions
Door	2	7′ × 7′
Door	4	3′ × 7′
Window	12	5′ × 6′
Window	4	3′ × 3′

6. Add the following items to the estimate in Exercise:

   • Precast lintel, 10′ long, 6″ wide, 8″ high, prestressed, 8 each

   • Precast lintel, 12′ long, 8″ wide, 8″ high, prestressed, 2 each

   Remember that you need to create two separate estimates (one for items and one for assemblies) and then merge them.

7. What is the cost of providing and installing steel stairs, cement-filled metal pan and picket rail, 20 risers, with landing? Eight flights are needed for a building located in Ann Arbor, Michigan.

8. A bank in Aspen, Colorado, needs three teller windows, bulletproof glazing, 44″×60″; and one drive-through teller window, painted steel, 72″×40″. Estimate the total cost.

Chapter 15

Approximate Estimates

Approximate estimates—also called *preliminary, budget, ballpark, rough,* or *back-of-the-envelope* estimates—are different from detailed, final estimates. They are widely used in the construction industry for these purposes:

- Feasibility studies for owners or investors to determine whether the proposed project is within their financial capability.

- To compare alternative projects and their approximate costs.

- To compare alternative designs for the same project. For example, an industrial building may be built of reinforced concrete or steel, and a roof built of either wood or steel trusses.

- To facilitate financial arrangements for the project with banks, investors, or other resources.

- As preliminary project estimates in cases where the detailed design is not complete. If the contract is cost-plus-fee type, an approximate estimate gives both the owner and the contractor an idea about the total cost, especially if a guaranteed maximum price (GMP) is needed.

Unlike detailed estimates, approximate estimates are not derived from actual material takeoff. Instead, they are based on average actual costs for recent similar projects. When using such past data, it is important to make all necessary cost adjustments between past projects and the proposed project and to allow for cost escalation. These adjustments include, but are not limited to, project location, project size, type of soil (and, thus, foundation), level of finish, labor productivity, and prevailing economic conditions.

RSMeans Project Costs

Through 2015, RSMeans collected cost data on completed construction projects and published summary results for various building types in Section 50 17 00 – Square Foot Costs in RSMeans *Building Construction Cost Data*. This cost data was collected from a large number of projects, analyzed, categorized, and averaged. Costs were presented for nearly 60 building types in the 1/4 (25ᵗʰ percentile), median (50ᵗʰ percentile), and 3/4 (75ᵗʰ percentile) categories in a Cost per Square Foot format for some of the major building component/discipline categories as well as the Total Project Cost per Square Foot. This limited breakdown of project costs for some of the major building component/discipline categories was sporadic due to the vague nature of some of the project cost reports received.

RSMeans did not publish this Square Foot Costs information in 2016 while it sought more reliable sources for this type of data. RSMeans found a reliable source and published Section 50 17 00 – Square Foot Costs once again, for a smaller set of 28 building types (see Figure 15.1), but not in time to be published in the 2017 printed products. RSMeans did, however, post the data to a companion website whose 2017 URL was given to customers upon request. Later in 2017, RSMeans made the information available in its electronic estimating software: in the compact disc-based *Costworks* and the web-based RSMeans Online. This information is published once again by RSMeans in some of its 2018 printed products and in the 2018 electronic estimating software.

50 17 | Square Foot Costs

		50 17 00	S.F. Costs	UNIT	UNIT COSTS			% OF TOTAL			
					1/4	MEDIAN	3/4	1/4	MEDIAN	3/4	
01	0000	Auto Sales with Repair		S.F.							01
	0100	Architectural			95	114	115	55.50%	66%	67%	
	0200	Plumbing			8	9.35	11.15	5.15%	5.45%	6.80%	
	0300	Mechanical			11.55	15.30	15.80	6.75%	8.95%	10.15%	
	0400	Electrical			16.45	20.50	21	9.05%	11.70%	11.90%	
	0500	Total Project Costs			160	171	171				
02	0000	Banking Institutions		S.F.							02
	0100	Architectural			144	177	215	59%	65%	69%	
	0200	Plumbing			5.80	8.10	11.25	2.12%	3.39%	4.19%	
	0300	Mechanical			11.50	15.90	18.75	4.41%	5.10%	10.75%	
	0400	Electrical			28	34	52	10.45%	13.05%	15.90%	
	0500	Total Project Costs			239	268	330				
03	0000	Court House		S.F.							03
	0100	Architectural			104	160	160	58.50%	63%	63%	
	0200	Plumbing			10.20	27	27	4.03%	15.30%	15.30%	
	0300	Mechanical			22	22	22	8.70%	8.70%	8.70%	
	0400	Electrical			25	28	28	11.10%	14.25%	14.25%	
	0500	Total Project Costs			177	253	253				
04	0000	Data Centers		S.F.							04
	0100	Architectural			171	171	171	68%	68%	68%	
	0200	Plumbing			9.40	9.40	9.40	3.71%	3.71%	3.71%	
	0300	Mechanical			24	24	24	9.45%	9.45%	9.45%	
	0400	Electrical			22.50	22.50	22.50	9%	9%	9%	
	0500	Total Project Costs			252	252	252				
05	0000	Detention Centers		S.F.							05
	0100	Architectural			159	168	178	52%	53%	60.50%	
	0200	Plumbing			16.75	20.50	24.50	5.15%	7.10%	7.25%	
	0300	Mechanical			21.50	30.50	36.50	7.55%	9.50%	13.80%	
	0400	Electrical			35	41	53.50	10.90%	14.85%	17.95%	
	0500	Total Project Costs			268	283	335				
06	0000	Fire Stations		S.F.							06
	0100	Architectural			95.50	115	169	52%	56%	65%	
	0200	Plumbing			9.10	12.50	14.35	4.86%	5.60%	6.20%	
	0300	Mechanical			12.35	16.90	23	5.45%	7.60%	10.50%	
	0400	Electrical			21	25.50	28.50	8.20%	11.95%	14.40%	
	0500	Total Project Costs			191	214	287				

Figure 15.1

RSMeans Online estimating offers the Square Foot Estimator module with which users can perform conceptual estimating using RSMeans building models comprised of RSMeans Assemblies. The Assemblies are comprised of RSMeans Unit Cost Lines (items) whose material, labor, and equipment costs are updated each quarter. The RSMeans Square Foot Estimator module allows the user to choose a type of building to estimate anywhere in the country with current localized costs. The user can choose various building parameters via pull-down options such as project location, building type, project size in gross SF, number of stories, story height, building envelope material, type and material of the building's structural frame, and optional additives such as extra elevators, surveillance cameras, upgraded kitchen appliances, etc. Once a project is estimated and saved, a report can be generated that gives a detailed breakdown, in the UniFormat II numbering schema, that lists major building component/system categories along with all RSMeans assemblies under those categories. The report can be modified by swapping out default assemblies for more applicable assemblies that meet project parameters. If the user needs further customization, the estimate needs to be exported in Excel format where the user can make unit price changes plus any other necessary adjustments.

Using Online RSMeans Square Foot Estimator

First, choose the Square Foot Estimator tab at the top of the screen; you need to fill in the Estimate Header Information. Some of the information is not mandatory, but it is recommended to fill them all. Information boxes with a red asterisk have to be filled (Figure 15.2).

After that, move to Step 1: Building Type to choose the Building Type and Wall/Framing Type from drop-down menus. In Step 2, you need to define

Figure 15.2

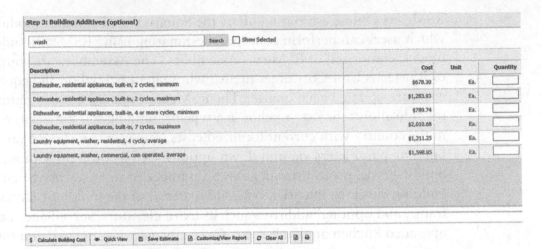

Figure 15.3

the building parameters. Note that the area (in square feet) represents the total, not per floor. Normal ranges are shown in yellow strips above some categories such as area, floors, and story height, although you can go out of that range. For number of stories, it is recommended to stay within the normal range, and if you need to go out of that range, you can choose a different building type (in Step 1).

Step 3: Building Additives is optional, and it has some additions such as appliances, closed-circuit surveillance, elevators, emergency lighting, laundry equipment, or smoke detectors. You can specify the quantity of any added item. You cannot adjust unit prices here but you can export the estimate to Excel, where you can adjust anything. You can do a word search for an item: In the screenshot (Figure 15.3) we did a word search on "wash" and we got six items, including dishwashers and washers. If you check the Show Selected box, you will see only selected items, i.e., those with quantities greater than zero. This is a good feature to zoom in on the items chosen.

Note that you can view the different steps in either List or Grid View (buttons on the upper right-hand side of the screen). Also, the top menu shows at the bottom of the screen for the user's convenience (see Figures 15.2 and 15.3) with these choices:

- Calculate Building Cost: After you entered/changed some information. Note that the Building Cost (blue box on the top of the screen) is not updated automatically. You need to click the Calculate Building Cost button if you would like to see the updated cost.

- Quick View: This will show a spreadsheet breaking down the estimate by type of work along with its percentage of the total cost.
- Save Estimate
- Customize/View Report
- Clear All
- Export to Excel
- Print

The default image shown for the building project is provided by RS Means Company, but you can remove the image or replace it with your own image.

Once you are done with Steps 1, 2, and 3, save the estimate. You can select Quick View to see the details (before or after saving the estimate). This shows you the assemblies the model is made of. You can click the Customize/View Report in order to customize your estimate by adding, swapping, or deleting assemblies (see Figures 15.4 and 15.5). More on using the Square Foot Estimator is shown in the following examples.

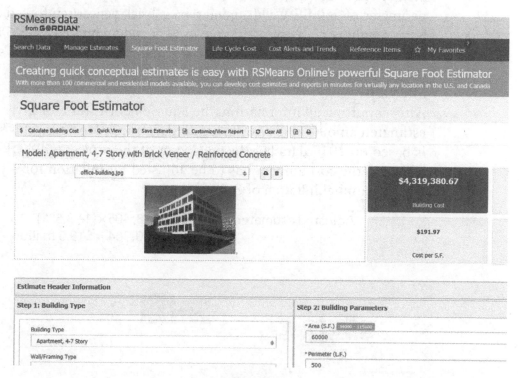

Figure 15.4

Figure 15.5

Example 1

Estimate the total cost for building a six-story office building in Chicago, Illinois. The project will start in April 2019 and is expected to finish in September 2020. The building will be rectangular with a cross-sectional area of 7,200 SF (which is the same for each of the six floors) and perimeter of 360 LF. Wall/framing type is brick veneer/reinforced concrete. Story height is 12', contractor's fee is 20% and architect's fee is 6%. Include a basement. Do not include any appliances or additional systems.

Solution

After inputting all information, the Square Foot Estimator gives us the estimated amount of $17,678,405 (Figure 15.6). However, this amount is based on 2017 data but the mid-point of the project will be January 2020. So the cost amount has to be adjusted for inflation for 3 years. Assume annual inflation of 3.5%:

$$\text{Adjusted estimated cost} = \$17,678,405 \times (1+3.5\%)^3$$
$$= \$19,600,364 \approx \$19.6 \text{ million}$$

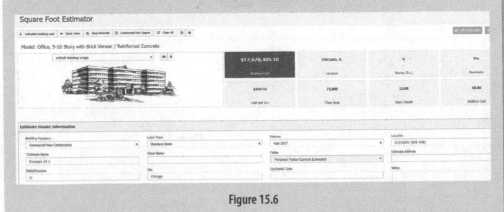

Figure 15.6

Example 2

Repeat example 1 with the following additions:

1. Closed-circuit television system (CCTV), surveillance, one station (camera & monitor), plus five additional camera stations.

2. Moving stairs, escalator type, 10′ ht, 32″ width, metal balustrade: Quantity = 2.

3. Metal detector, walk through portal type, single zone: Quantity = 2.

4. Uninterruptible power supply with standard battery pack, 15 kVA/12.75 kW: Quantity = 1.

Solution

Going back to the estimate and adding these items results in a cost estimate of $18,139,622. With adjusting to inflation it becomes:

$$\text{Adjusted estimated cost} = \$18,139,622 \times (1 + 3.5\%)^3$$
$$= \$20,111,723 \approx \$20.1\,\text{million}$$

Example 3

Estimate the total cost for building a movie theater in Miami, Florida. The project will start in November 2019 and is expected to finish in October 2020. The building will have rectangular cross section with a cross-sectional area of 15,000 SF and perimeter of 520 LF. Wall/framing type is precast concrete/steel joists and story height is 20′. Contractor's fee is 18 percent and architect's fee is 7 percent. Include 12 emergency lighting (25 watt, battery-operated nickel cadmium), 750 upholstered spring seats, 20 smoke detector (ceiling type), 4 amplifiers 250 Watt, and 8 speakers. Assume 4 percent annual inflation (Figure 15.7).

Solution

After inputting all information, the Square Foot Estimator gives us the estimated amount of $2,747,742:

$$\text{Adjusted estimated cost} = \$2,747,742 \times (1 + 4\%)^3$$
$$= \$3,090,836 \approx \$3.09\,\text{million}$$

The Quick View allows us to see the breakdown of the cost estimate. We can also export to Excel, where we can see the breakdown and do any unit price or other adjustments.

RSMeans data from **GORDIAN** Square Foot Cost Estimate Report   Date: 3/20/2018

Estimate Name:	Example 15-3
	Miami , FL
	Movie Theater with Precast Concrete / Steel
Building Type:	Joists
Location:	MIAMI, FL
Story Count:	1
Story Height (L.F.):	20
Floor Area (S.F.):	15000
Labor Type:	STD
Basement Included:	No
Data Release:	Year 2017
Cost Per Square Foot:	$183.18
Building Cost:	$2,747,741.73

Costs are derived from a building model with basic components.

Scope differences and market conditions can cause costs to vary significantly.

		% of Total	Cost Per S.F.	Cost
A Substructure		5.49%	7.96	119,445.22
A1010	**Standard Foundations**		3.4	51,001.42
	thick		2.24	33,625.28
	KSF, 12" deep x 24" wide		1.11	16,666.00
	- 0" square x 12" deep		0.05	710.14
A1030	**Slab on Grade**		4.34	65,062.80
	Slab on grade, 4" thick, non industrial, reinforced		4.34	65,062.80
A2010	**Basement Excavation**		0.23	3,381.00
	site storage		0.23	3,381.00
B Shell		37.10%	53.83	807,443.18
B1010	**Floor Construction**		1.69	25,304.81
	span, 22.5" deep, 100 PSF superimposed load, 145 PSF total load		1.69	25,304.81
B1020	**Roof Construction**		10.71	160,649.70
	wall, 60'x50' bay, 40 PSF superimposed load, 71" deep, 65 PSF total load		10.16	152,385.15
	wall, 60'x50' bay, 40 PSF superimposed load, 71" deep, 65 PSF total load,		0.55	8,264.55
B2010	**Exterior Walls**		28.23	423,512.63
	insulation, low rise		28.23	423,512.63
B2020	**Exterior Windows**		6.7	100,512.77
	intermediate horizontals		2.76	41,425.07
	Glazing panel, insulating, 1" thick units, 2 lites, 1/4" float glass, clear		3.94	59,087.70
B2030	**Exterior Doors**		1.13	16,897.66
	hardware, 6'-0" x 10'-0" opening		0.4	6,043.78
	0" opening		0.72	10,853.88
B3010	**Roof Coverings**		5.37	80,565.61
	mopped		2.36	35,369.10
	Insulation, rigid, roof deck, composite with 2" EPS, 1" perlite		1.84	27,635.10
	Roof edges, aluminum, duranodic, .050" thick, 6" face		0.81	12,219.09
	Gravel stop, aluminum, extruded, 4", mill finish, .050" thick		0.36	5,342.32
C Interiors		14.75%	21.4	321,036.46
C1010	**Partitions**		2.9	43,557.60
	Concrete block (CMU) partition, light weight, hollow, 6" thick, no finish		2.9	43,557.60
C1020	**Interior Doors**		1.5	22,570.83
	3'-0" x 7'-0" x 1-3/8"		1.5	22,570.83
C1030	**Fittings**		0.46	6,892.59
	Toilet partitions, cubicles, ceiling hung, stainless steel		0.37	5,479.40
	Directory boards, outdoor, 36" x 36"		0.09	1,413.19
C2010	**Stair Construction**		3.04	45,573.25
	Stairs, steel, pan tread for conc in-fill, picket rail,20 risers w/ landing		3.04	45,573.25
C3010	**Wall Finishes**		2.33	34,944.55
	2 coats paint on masonry with block filler		1.28	19,259.47
	Painting, masonry or concrete, latex, brushwork, primer & 2 coats		1.05	15,685.08
C3020	**Floor Finishes**		5.68	85,250.69
	Carpet, tufted, nylon, roll goods, 12' wide, 36 oz		4.32	64,808.29
	Carpet, padding, add to above, 2.7 density		0.89	13,360.37
	Tile, ceramic natural clay		0.47	7,082.03
C3030	**Ceiling Finishes**		5.48	82,246.95
	channel grid, suspended support		5.48	82,246.95

Figure 15.7

D Services		22.31%	32.36	485,462.14
D2010	**Plumbing Fixtures**		**8.9**	**133,517.88**
	Water closet, vitreous china, bowl only with flush valve, wall hung		6.44	96,589.92
	Urinal, vitreous china, wall hung		0.47	6,978.90
	Lavatory w/trim, vanity top, PE on CI, 19" x 16" oval		0.72	10,836.42
	Service sink w/trim, PE on CI, wall hung w/rim guard, 22" x 18"		0.4	6,057.53
	Water cooler, electric, wall hung, dual height, 14.3 GPH		0.87	13,055.11
D2020	**Domestic Water Distribution**		**0.42**	**6,365.82**
	Gas fired water heater, residential, 100< F rise, 30 gal tank, 32 GPH		0.42	6,365.82
D2040	**Rain Water Drainage**		**1.05**	**15,720.93**
	Roof drain, DWV PVC, 5" diam, 10' high		0.53	7,988.90
	Roof drain, DWV PVC, 5" diam, for each additional foot add		0.52	7,732.03
D3050	**Terminal & Package Units**		**8.83**	**132,475.50**
	38.33 ton		8.83	132,475.50
D4010	**Sprinklers**		**3.48**	**52,232.22**
	Wet pipe sprinkler systems, steel, light hazard, 1 floor, 10,000 SF		2.94	44,162.55
	SF		0.54	8,069.67
D4020	**Standpipes**		**0.82**	**12,373.19**
	Wet standpipe risers, class III, steel, black, sch 40, 4" diam pipe, 1 floor		0.67	10,058.63
	floors		0.15	2,314.56
D5010	**Electrical Service/Distribution**		**1.23**	**18,454.28**
	wire, 3 phase, 4 wire, 120/208 V, 400 A		0.33	5,011.30
	Feeder installation 600 V, including RGS conduit and XHHW wire, 400 A		0.53	7,957.60
	V, 3 phase, 400 A		0.37	5,485.38
D5020	**Lighting and Branch Wiring**		**5.78**	**86,660.70**
	Receptacles incl plate, box, conduit, wire, 2.5 per 1000 SF, .3 watts per SF		1.54	23,148.30
	Miscellaneous power, to .5 watts		0.12	1,873.20
	Central air conditioning power, 3 watts		0.52	7,866.60
	fixtures @32 watt per 1000 SF		2.26	33,847.80
	Daylight dimming control system, 10 fixtures per 1000 SF		1.33	19,924.80
D5030	**Communications and Security**		**1.68**	**25,127.55**
	detectors, includes outlets, boxes, conduit and wire		0.9	13,498.80
	Fire alarm command center, addressable with voice, excl. wire & conduit		0.78	11,628.75
D5090	**Other Electrical Systems**		**0.17**	**2,534.07**
	gas/gasoline operated, 3 phase, 4 wire, 277/480 V, 7.5 kW		0.17	2,534.07
E Equipment & Furnishings		20.35%	29.52	442,869.72
E1020	**Institutional Equipment**		**7.57**	**113,527.96**
	platters & autownd, econ		0.39	5,875.00
	rectifiers, xenon, 1000W		0.53	7,974.24
	mm, economy		0.89	13,300.00
	wall, acrylic, 1/4" thick		5.76	86,378.72
E1090	**Other Equipment**		**17.35**	**260,260.38**
	sealed beam light, 25 W, 6 V each		0.36	5,386.44
	750.00-Auditorium chair, fully upholstered, spring seat		16.06	240,952.50
	wires & conduit		0.32	4,833.40
	conduits		0.5	7,551.00
	conduits		0.1	1,537.04
E2010	**Fixed Furnishings**		**4.61**	**69,081.38**
	upholstered, economy		4.61	69,081.38
F Special Construction		0%	0	0
G Building Sitework		0%	0	0
SubTotal		100%	$145.08	$2,176,256.72
Contractor Fees (General Conditions,Overhead,Profit)		18.00%	$26.12	$391,726.21
Architectural Fees		7.00%	$11.98	$179,758.81
User Fees		0.00%	$0.00	$0.00
Total Building Cost			$183.18	$2,747,741.73

✏ ** Indicates Assemblies or Components have been customized.

Inflation period	3.0 years
Inflation rate	4%
Adjusted Cost	$3,090,836

Figure 15.7 (continued)

RSMeans Project Costs

If we would like to use the RSMeans Square Foot Costs (Figure 15.1), let's see what the information includes:

- *Cost per square foot for a variety of building types, sizes, and cost ranges.* Each project type is arranged in order of square foot cost, from the least expensive to the most expensive. The cost in the middle of the set (not necessarily the average) is reported as the median. The ¼ cost represents the 25th percentile, which means it is higher than 25 percent of similar projects but lower than the other 75 percent of similar projects; thus, it is on the lower side. The ¾ cost represents the 75th percentile, which means it is higher than 75 percent of similar projects but lower than the other 25 percent of similar projects; thus, it is on the higher side. The three estimates (¼, median, and ¾) are used for below-average, average, and above-average projects costs.

  It is interesting to observe that the difference between the ¼ cost and the median cost is almost always smaller than the difference between the median cost and the ¾ cost. This is because the choices are limited when we like to cut the cost, but the sky is the limit when we go in the luxury direction.

- *Cost per square foot for different types of work.* Types of work include site work, masonry, equipment, plumbing, HVAC, and electrical, given in dollar amounts and as a percentage of the total cost. The ¼, median, and ¾ percentages of total cost represent the range of total contract cost for that portion of work. Caution is recommended, as the "% of Total" figures are not directly related to the square foot costs, because they are not based on identical data sets. A median project may involve a higher-than-average percentage of its total cost for mechanical and/ or electrical for some reason, such as harsh weather. In general, it is better to use the unit cost ($/SF) rather than the percentage of total.

When using RSMeans Project Costs to find the approximate estimate for a project, follow these guidelines:

1. Pick the title from the Project Costs that matches your project type.

2. Pick the suitable unit cost (e.g., ¼, median, and ¾), according to whether the cost of the proposed project is below average, average, or above average. In some cases, depending on the user's discretion, a cost figure between the ¼ cost and the median may be taken for a project slightly below average; similarly, a cost figure between the median and the ¾ figure may be taken for a project slightly above average. The user may also choose a figure slightly below the ¼, or higher than the ¾, depending on his or her discretion.

3. Apply a factor of scale by comparing the proposed building size to the typical size for similar projects, found in the Square Foot Project Size Modifier, Figure 15.8.

Square Foot Project Size Modifier

One factor that affects the S.F. cost of a particular building is the size. In general, for buildings built to the same specifications in the same locality, the larger building will have the lower S.F. cost. This is due mainly to the decreasing contribution of the exterior walls plus the economy of scale usually achievable in larger buildings. The Area Conversion Scale shown below will give a factor to convert costs for the typical size building to an adjusted cost for the particular project.

The Square Foot Base Size lists the median costs, most typical project size in our accumulated data and the range in size of the projects.

The Size Factor for your project is determined by dividing your project area in S.F. by the typical project size for the particular Building Type. With this factor, enter the Area Conversion Scale at the appropriate Size Factor and determine the appropriate cost multiplier for your building size.

Example: Determine the cost per S.F. for a 152,600 S.F. Multi-family housing.

$$\frac{\text{Proposed building area} = 152,600 \text{ S.F.}}{\text{Typical size from below} = 76,300 \text{ S.F.}} = 2.00$$

Enter Area Conversion scale at 2.0, intersect curve, read horizontally the appropriate cost multiplier of .94. Size adjusted cost becomes .94 x $187.00 = $175.78 based on national average costs.

Note: For Size Factors less than .50, the Cost Multiplier is 1.1
For Size Factors greater than 3.5, the Cost Multiplier is .90

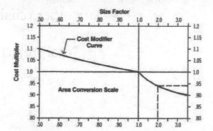

System	Median Cost (Total Project Costs)	Typical Size Gross S.F. (Median of Projects)	Typical Range (Low – High) (Projects)
Auto Sales with Repair	$171.00	25,300	8,200 – 28,700
Banking Institutions	268.00	26,300	3,300 – 28,700
Detention Centers	283.00	42,000	12,300 – 183,300
Fire Stations	214.00	14,600	6,300 – 29,600
Hospitals	360.00	137,500	54,700 – 410,300
Industrial Buildings	$93.50	16,900	5,100 – 200,600
Medical Clinics & Offices	212.00	5,500	2,600 – 327,000
Mixed Use	194.00	49,900	14,400 – 49,900
Multi-Family Housing	187.00	76,300	12,500 – 1,161,500
Nursing Home & Assisted Living	125.00	16,200	1,500 – 242,600
Office Buildings	176.00	10,000	1,100 – 930,000
Parking Garage	42.00	174,600	99,900 – 287,000
Parking Garage/Mixed Use	157.00	5,300	5,300 – 318,000
Police Stations	273.00	15,400	15,400 – 31,600
Public Assembly Buildings	253.00	30,500	2,200 – 235,300
Recreational	275.00	2,300	1,500 – 223,800
Restaurants	273.00	6,100	5,500 – 42,000
Retail	102.00	28,700	5,800 – 61,000
Schools	204.00	30,000	5,500 – 410,800
University, College & Private School Classroom & Admin Buildings	261.00	89,200	9,400 – 196,200
University, College & Private School Dormitories	204.00	50,800	1,500 – 126,900
University, College & Private School Science, Eng. & Lab Buildings	262.00	39,800	36,000 – 117,600
Warehouses	113.00	2,100	600 – 303,800

Figure 15.8 Square Foot Project Size Modifier

Use the curve to estimate the cost multiplier. First, locate the size factor value at the X-axis (top or bottom), and then use a ruler and a pencil to draw a thin vertical line until it intersects with the curve. From the point of intersection, draw a horizontal line until it intersects with the Y-axis (right or left). That number is your cost multiplier. Multiply the proposed building cost by the cost multiplier to determine the adjusted cost for the project.

4. Determine the city cost index (CCI) for the proposed project location. Multiply the "weighted average" by the adjusted project cost. This factor represents the balanced adjustment factor for all project costs, located at the bottom of the detailed CCI division factors.

5. Determine the effect of future cost escalation on the project. Calculate the approximate time period in years, n, between January 2017 (the RSMeans data date) and the midpoint of the project. Estimate the annual inflation, i, for the next few years. You may use:

- Your own educated forecast
- RSMeans Historical Cost Indexes to extrapolate using the past 5, 10, or 20 years (remember that future inflation may not necessarily behave in the same manner as past inflation).
- Any other reliable source for economic forecast

$$\text{Total project cost} = \text{Project size (SF)} \times \text{Unit cost}\left(\frac{\$}{SF}\right)$$

$$\times \text{Cost multiplier} \times CCI \times (1+i)^n$$

Example 4

Estimate the total cost for a high-end 7,000-SF medical office building in Chicago, Illinois. The building will be built during 2019.

Solution

Use RSMeans item 50 17 00.10 05000, the ¾ (75th percentile) unit cost: $265/SF.
From the Square Foot Project Size Modifier (Figure 15.2), the typical size for a bank is 5,500 SF.

$$\text{Size factor} = 7,000/5,500 = 1.27$$

Using the Area Conversion Scale in the Project Size Modifier, we find the cost multiplier = 0.97.
CCI for Chicago, Illinois, weighted average, total = 120 percent.
Assume annual inflation = 4 percent. Assume the project midpoint is July 2019; then $n = 2\frac{1}{2}$ years.

$$\text{Total cost} = 7,000 \text{ SF} \times \$265/\text{SF} \times 0.97 \times 120\% \times (1+0.04)^{2.5}$$

$$= \$2,381,663$$

$$\approx \$2,382,000$$

Example 5

If the electrical and mechanical work in the medical office building project (Example 1) will be subcontracted, how much do you estimate that subcontract will be?

Solution

From line 50 17 00.10 0300 and 0400, the total mechanical and electrical cost is $43.20 (18.70+24.50)/SF for the median project and $78.50/SF for the ¾ (75th percentile) project. As a percentage of the total, the median is 22.55% and the ¾ (75th percentile) is 30.95%.

Since we already took the ¾ (75th percentile) for the total cost, let's take the median percentage of that cost, 22.55%.

The decision to use the median, the ¼, or the ¾ number depends on the design (type of system), which, in turn, depends on several factors such as the class (level of luxury) of the building and climatic conditions at the location of the proposed project. For example, a building in Phoenix, Arizona, is expected to require a bigger air conditioning system than a similar building in Los Angeles, California.

$$\text{Total cost for mechanical and electrical} =$$
$$\$2,381,663 \times 22.55\% = \$537,065 \approx \$537,000$$

If we like to use the number (not the percentage) for total mechanical and electrical, let's use a number between the median and ¾: $21.6 for mechanical and $39.25 for electrical. CCI for Chicago for mechanical is 117.5 percent and for electrical is 119.1 percent.

$$\text{Total mechanical cost} = 7,000\,\text{SF} \times \$21.60/\text{SF} \times 0.97 \times 117.5\% \times (1+0.04)^{2.5}$$
$$= \$190,084$$
$$\text{Total electrical cost} = 7,000\,\text{SF} \times \$39.25/\text{SF} \times 0.97 \times 119.1\% \times (1+0.04)^{2.5}$$
$$= \$328,945$$
$$\text{Total mechanical and electrical cost} = \$519,028 \approx \$519,000$$

The two numbers, $537,000 and $519,000, are relatively close. Remember that this is an approximate estimate without knowing the specifications.

Keep in mind that using the median percentage does not mean we are using the median cost because we are multiplying this median percentage by the total, based on above average (¾) total cost.

Example 6

Calculate the approximate cost for a five-story apartment building that contains 40 apartments, each with 780-SF area and building efficiency ratio of 75 percent.[1] The project will be built in Los Angeles, California, between September 2019 and May 2020. Find also the cost per unit (apartment).

Solution

Use RSMeans item 50 17 00.12 0500 Since there was no explicit or implicit indication as to whether this complex is average, below, or above average, assume it is average.

---

1. The building efficiency ratio is the proportion of a building's net (leasable) area to the gross area. The difference between gross and net areas includes hallways, stairways, and other common areas inside the building.

$$\text{Total net area} = 40 \text{ apartments} \times 780 \text{ SF each} = 31{,}200 \text{ SF}$$

$$\text{Total building (gross) area} = 31{,}200 \,/\, 0.75 = 41{,}600 \text{ SF}$$

Use line 50 17 00.12 0500: the median cost is $187/SF.
Size factor = 41,600/76,300 = 0.55. Then, using the Area Conversion
Scale in the Project Size Modifier; the cost multiplier = 1.08.
CCI, Los Angeles, California, weighted average, total, is 113.4 percent.
Assume 4 percent annual inflation. Time period between RSMeans data
(January 2017) and project midpoint is 3 years.

$$\text{Total cost} = 41{,}600 \times \$187\,/\,\text{SF} \times 1.08 \times 113.4\% \times (1 + 0.04)^3$$
$$\approx \$10{,}716{,}964 \approx \$10{,}717{,}000$$

$$\text{Cost per apartment} = \$10{,}716{,}964 \,/\, 40 = \$267{,}924 \approx \$268{,}000$$

Example 7

Repeat Example 6 using the Online Square Foot Estimator (Figure 15.9).
Assume story height = 11 feet.

Solution:

We entered all information except for Step 3 (Building Additives).

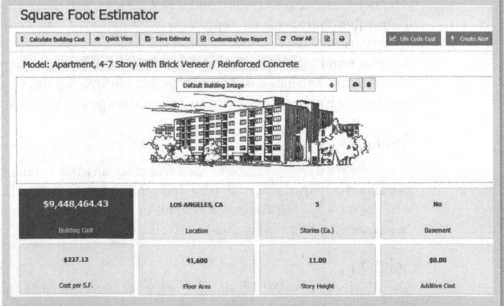

Figure 15.9

The estimated total is $9,448,464, but this is in 2017 dollars. If we
apply the expected inflation, the answer will be $10,628,000.

Chapter 15 Exercises—Set A

For solutions to the exercises in Set A, see the link to the student companion website at www.wiley.com/go/constructionestim5e.
Note: Use the Online Square Foot Estimator for Exercises 1 through 11.
Unless otherwise mentioned, assume a 4 percent expected annual inflation.

1. A large discount retail company is planning to build an 80,000 SF giant supermarket in Atlanta, Georgia. The work should start in September 2019 and be completed by March 2021. What do you expect the total cost to be? Keep program default values as is and make any necessary assumptions. Do not add any building additives.

2. The mechanical and electrical work in Exercise 1 will be subcontracted. What is your estimate for that subcontract?

3. Repeat Exercise 1 assuming the contractor fee to be only 20%, the architectural fee is 5%, and the project to be "green."

4. Jails in the state of Texas are overflowing with inmates. The state has decided to build a new, low-cost 90,000 SF jail just outside Houston. Estimate the approximate total cost of the new jail, knowing that construction will start in January 2019 and will take one year.

5. Calculate the costs for Exercise 4 for:

   A. Plumbing

   B. HVAC

   C. Electrical

6. In Exercise 5, add:

   • A camera and monitor with 10 additional camera stations.

   • Two 2,500# elevators

   • Four emergency generators, complete systems: two gas 70 KW and two diesel 150 KW

   • Two flagpoles, aluminum, 40′ high

   • Sound system: one amplifier, 250 watts, and 8 wall speakers

7. Estimate the approximate cost of building a two-story fire station in Tuscan, Arizona, with a total area of 12,500 SF and face brick & concrete block/precast concrete. Construction is expected to start in August 2019 and will take 15 months to complete.

8. What would be the cost of the fire station in Exercise 7 if it has face brick & concrete block/steel joists?

9. How much would the fire station cost in Exercise 7 if the project is green?

10. A private elementary school is needed in Huntsville, Alabama, with a total area of 48,000 SF and the walls will be tilt-up concrete panels/reinforced concrete. The school should be ready for the 2020–2021 year. (Construction takes about six months.) How much will the total cost be?

11. Add the following items to the estimate in Exercise 10:

Item	Quantity
Master time clock system, master controller, clocks & bells, 20 room, excl. wires & conduits	1
Emergency lighting units, nickel cadmium battery operated, twin sealed beam light, 25 W, 6 V each	4
Flagpoles, fiberglass, tapered, ground set, 23' high, excludes base or foundation	2
Broiler, commercial kitchen equipment, without oven, standard	1
Cooler, commercial kitchen equipment, reach-in, beverage, 6' long	1
Dishwasher, commercial kitchen equipment, 10 to 12 racks per hour	2
Food warmer, commercial kitchen equipment, counter, 1.2 KW	1
Freezers, commercial kitchen equipment, reach-in, 44 CF	1
Ice-cube maker, commercial kitchen equipment, 50 lbs per day	1
Range, commercial kitchen equipment, restaurant type, 6 burners & 1 standard oven, 36" wide $3	1
Lockers, steel, baked enamel, double tier, set up, 60" or 72"	200

Use the Square Foot Costs (printed materials[2]) to solve the following exercises. *If no hint is given, the median unit cost should be used.* Clearly state any assumptions you make:

12. Repeat Exercise 4.

13. Estimate the total cost for building a luxury six-story apartment building, with 24 apartments that average 960 SF. Add 5,700 SF for common areas. The midpoint of construction, in New York City, New York, will be summer of 2020.

14. The mechanical and electrical work in Exercise 13 will be subcontracted. What is your estimate for that subcontract?

15. Redo Exercise 13 if the complex is "slightly above average."

16. Estimate the total cost for building a 160,000 SF hospital in Columbia, South Carolina. Provide two numbers: current (2017) prices and three years later.

---

2. Printed materials can be found in RSMeans Online estimating by selecting Reference Items->Square Foot->Project Cost Square Foot Data->Square Foot Cost

Chapter 15 Exercises—Set B

Solutions to Set B exercises are available to instructors only; see the link to the instructor's companion website at www.wiley.com/go/ constructionestim5e.

Note: Use the Online Square Foot Estimator for Exercises 1 through 9 and the Square Foot Costs (printed materials) for Exercises 10-16. If no hint is given, the median unit cost should be used. Unless otherwise stated, assume a 4 per cent expected annual inflation. Clearly state any assumptions you make:

1. A membership discount store is going to put a new store (use warehouse) in Los Angeles, California. Construction will be during the second half of 2019. The store will cover 40,000 SF. How much do you expect it to cost?

2. The mechanical and electrical work in Exercise 1 will be subcontracted. How much do you estimate that subcontract to be?

3. An auto sales garage is to be built in Des Moines, Iowa, between March and November 2020, with a total area of 22,000 SF. Use tilt-up concrete panels/rigid steel system. How much do you expect it to cost?

4. How much would the project in Exercise 3 cost if we use stucco and concrete block/bearing walls system?

5. Estimate the cost of a six-story college dormitory with identical 12,000 SF floor plan (per story) in Richmond, Virginia. Wall/Framing type is brick veneer/rigid steel. The project will start in May 2020 and finish in 14 months. Contractor's fee is 22% and architect's fee is 5.5%. Include a basement.

6. Calculate the costs for Exercise 5 for:

   A. Plumbing

   B. HVAC

   C. Electrical

7. Estimate the approximate cost of building a one-story candy factory in Panama City, Florida, with a total area of 17,500 SF. The owner is on a tight budget and will try to cut corners as much as possible. Construction is expected to start immediately.

8. What would the cost of the factory in Exercise 7 be if the construction will be postponed two years?

9. How much would a factory, similar to the one in Exercise 7, cost if is to be built in New York City?

Use the Square Foot Costs (printed materials) for following exercises:

10. Estimate the total cost for a 3-story parking garage in Lexington, Kentucky, with a total area 63,000 SF. Construction will be in mid-2019.

11. A court house is needed in Vancouver, Canada. It will have a total area of 30,000 SF. The building should be ready by the end of July 2021. (Construction takes an average of one year.) How much will the total cost be?

12. How much will the plumbing and mechanical in Exercise 11 cost? (Figure each separately.)

13. How much would a court house similar to that in Exercise 11 cost if it were built at the same time in San Francisco, California?

14. Estimate the total cost for erecting a 5-story multi-family housing building. It will have 50 apartments averaging 1,000 SF each, with a building efficiency ratio of 80 percent, and is to be constructed in New York City, New York, during 2019.

15. The mechanical and electrical work in Exercise 14 will be subcontracted. What is your estimate for that subcontract?

16. Redo Exercise 14 as if the building were to be built in Indianapolis, Indiana.

Appendix A
Contractor's Cash Flow

Time Value of Money

The equations assume compound interest.

P: Value or amount of money at a time designated as the present or time 0. Also, *P* is referred to as present worth (PW), present value (PV), net present value (NPV), discounted cash flow (DCF), and capitalized cost (CC).

F: Value or amount of money at some future time. Also, *F* is called future worth (FW) and future value (FV).

A: Series of consecutive, equal, end-of-period amounts of money. Also, *A* is called the annual worth (AW) and equivalent uniform annual worth (EUAW).

n: Number of interest periods

i: Interest rate or rate of return per time period

APR: Annual percentage rate

Single Payment

$$F = P \times (1 + i)^n \qquad (Equation\ 1)$$

Equal Uniform Payments

$$F = A\left[\frac{(1+i)^n - 1}{i}\right] \qquad (Equation\ 2)$$

$$P = A\left[\frac{(1+i)^n - 1}{i(1+i)^n}\right] \qquad (Equation\ 3)$$

$$A = F\left[\frac{i}{(1+i)^n - 1}\right] \qquad (Equation\ 4)$$

$$A = P\left[\frac{i(1+i)^n}{(1+i)^n - 1}\right] \qquad (Equation\ 5)$$

Readers also can calculate interest by using interest tables, business and scientific calculators, or computer programs. Microsoft Excel has an excellent but often-overlooked template called *Loan Amortization* that comes installed with the software. It is very useful in calculating interest, particularly in financing loans with equal payments, such as homes and cars.

Progress Payment Process and the Contractor's Cash Flow

Construction projects usually take months or years. The contractor has to spend money even before construction begins, and continues spending throughout the execution of the project. Traditionally, the contractor submits a *payment request*[1] to the owner demanding payment for materials already installed and services already rendered. The owner—or its representative—verifies item quantities claimed by the contractor, their unit prices as compared to the schedule of values,[2] and the calculations, and then take an action: approve, approve with exceptions, or reject.

This approval process can take two to four weeks or even more but is usually restricted by time limit specified in the contract. Even for approved amounts, owners usually deduct a percentage called *retainage* and hold it until the completion of the project.[3] This entire process is called a *progress payment*.

Subcontractors follow the same process, but they have to submit their requests to the general contractor with a deadline earlier by two or three business days than the contractor's own deadline to the owner. Once the owner pays the general contractor, the general contractor pays its subcontractors. Owners usually require lien releases (waivers) from the general contractor and all subs and vendors for items paid for in every progress payment. Lien releases protect the owner from possible liens against its property in case a sub or vendor did not receive its payment from the general contractor.

The progress payment process puts the contractor's cash flow in a negative mode from the very beginning until the final payment for two reasons: the delay between the spending and reimbursement in every cycle and the owner's withholding of the retainage. This fact underscores the importance of financial management. The contractor must forecast the cash flow diagram for several reasons, but two of the most important reasons are listed next:

1. To calculate the cost of borrowing money throughout the project execution (construction) phase. Even if the contractor has enough cash and does not need to borrow money, the cost of interest has to be added as a lost opportunity for investing the money.

2. To calculate the credit line (or cash on hand) needed to match or exceed the maximum debt (negative cash flow) expected during the project.

---

1. Also called *pay request*, *payment requisition*, or *pay requisition*.

2. Defined in Chapter 3.

3. Retainage is a portion, usually 10 percent, of the eligible progress payment, that is held by the owner until the contractor fulfills its contractual obligations. The contract usually specifies the amount and conditions of the retainage. In large projects, it is customary to either reduce the retainage percentage or stop retaining any money by the owner after the project reaches certain completion stage, such as 50 percent complete. Retainage is also called *retention*.

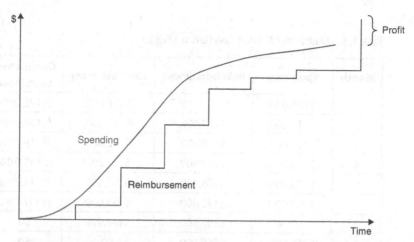

Figure A.1   Typical Contractor's Spending and Reimbursement in a Project

Example 1

A contractor is planning to start a construction project that is expected to take six months to complete. Its spending will be spread out over the six months, as shown in Table A.1.

Table A.1   Contractor's Monthly Spending in a Project

Month	1	2	3	4	5	6	Total
Spending	$126,000	$144,000	$192,000	$185,000	$150,000	$68,000	$865,000

The contractor desires a profit of 5 percent of its total cost of the project. The retainage is 10 percent, and the owner reimburses the contractor two weeks after the submission of the monthly pay request (progress payments). Interest (annual percentage rate [APR]) is 12 percent, compounded monthly. What is the minimum line of credit the contractor needs? What should the total bid of the contractor be?

Solution

In this example, a simplifying assumption is made. We assume a one-month lag between contractor's spending and owner's reimbursement, as shown in Figure A.2. This assumption comes from Table A.2.

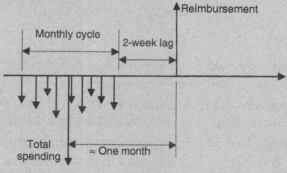

Figure A.2   Contractor's Progress Payment Monthly Cycle

281

Table A.2   Contractor's Monthly Cash Flow in a Project

Month	Spending	Reimbursement	Net Cash Flow	Cumulative Cash Flow	Future Value
1	($126,000)	0	($126,000)	($126,000)	($133,751.54)
2	($144,000)	$113,400	($ 30,600)	($156,600)	($ 32,160.91)
3	($192,000)	$129,600	($ 62,400)	($219,000)	($ 64,933.69)
4	($185,000)	$172,800	($ 12,200)	**($231,200)**	($ 12,569.67)
5	($150,000)	$166,500	$ 16,500	($214,700)	$ 16,831.65
6	($ 68,000)	$135,000	$ 67,000	($147,700)	$ 67,670.00
7		$147,700	$147,700	$0	$147,700.00
Total	($865,000)	$865,000	$0	$0	($ 11,214.16)

Explanation of the numbers in the table: Amounts in the "Spending" column represent the contractor's expected monthly spending, aggregated as one payment in the middle of the cycle, as shown in Figure A.2. Amounts in the "Reimbursement" column represent 90 percent of the previous month's spending. This is because owner retains 10 percent of the money and it takes two weeks from the end of the cycle, or roughly one month between the midpoint of spending, again as shown in Figure A.2. Amounts in the "Net Cash Flow" column are simply the algebraic summation of the previous two columns. The number in the first cell of the "Cumulative Cash Flow" column is equal to the cell in the same row of the "Net Cash Flow" column. Numbers in next cells are equal to the amount in the two cells: upper and left (previous cumulative plus current net). Finally, amounts in the "Future Value" column represent the future value of those in "Net Cash Flow" column, where "future" is defined as the point when final payment is made (End of the project + 2 weeks, or "month 7"), using equation 1 (single payment). Amounts in parentheses are negative.

So:
$$F_1 = -126,000 \times (1+0.01)^6 = -\$133,751.54$$
$$F_2 = -30,600 \times (1+0.01)^5 = -\$32,160.91$$
And so on until $F_7 = \$147,700 \times (1+0.01)^0 = \$147,700$
$$\Sigma F_i = -\$11,214$$

To answer the two questions of the example:
The minimum line of credit the contractor needs = $231,200, which is the maximum debt the contractor expects. The contractor expects this will happen at the end of the fourth month, which is the midpoint of the project. In reality, this is generally true because contractor spending usually follows an S-curve (see Figure A.1): It starts slow, picks up, and

increases toward the middle of the project and then tapers off toward the end. So the cumulative cash flow keeps increasing (in absolute value) while spending is more than reimbursement. Once we reach the point when spending is decreasing so the reimbursement (for last month's spending) is larger than the spending, then the contractor's debt starts decreasing.

The total bid of the contractor should be:

Total expenses (direct and indirect)	$865,000
Profit @ 5%	43,250
Interest	11,214
Total bid	$919,464
	≈ $920,000

This is a relatively small project, in both cost and duration. The interest was estimated at $11,214, which is roughly 1.3 percent of the total expenses. Also, we calculated that the contractor will at some point be in debt by $231,000. Things can get much more complicated with multimillion-dollar projects and duration of years. This result shows the importance of the cash flow analysis on the profitability and continuous operations of the construction company. In some contracts, owners' reimbursement time is longer than two weeks. The calculation of interest will have to be adjusted accordingly.

Please see the "Cash-Flow" spreadsheet (in Excel format) and cash-flow diagrams included on the book companion site for this solution.

Example 2

Repeat Example 1 but allow the owner six (instead of two) weeks to reimburse the contractor.

Solution

In this case, the interest on the payments will be two months:

Month	Spending	Reimbursement	Net Cash Flow	Cumulative Cash Flow	Future Value
1	($126,000)	$0	($126,000)	($126,000)	($135,089)
2	($144,000)	$0	($144,000)	($270,000)	($152,859)
3	($192,000)	$113,400	($ 78,600)	($348,600)	($ 82,609)
4	($185,000)	$129,600	($ 55,400)	($404,000)	($ 57,649)
5	($150,000)	$172,800	$ 22,800	($381,200)	$ 23,491
6	($ 68,000)	$166,500	$ 98,500	($282,700)	$100,480
		$135,000	$135,000	($147,700)	$136,350
		$147,700	$147,700	$0	$147,700
Total	($865,000)	$865,000	$0		($ 20,186)

The total bid of the contractor should be:

Total expenses (direct and indirect)	$865,000
Profit @ 5%	43,250
Interest	20,186
Total bid	$928,436
	≈ $930,000

This minor change in the payment process added about $9,000 to the contractor's cost, which should be taken in consideration when preparing the cost estimate. What is also important is that the contractor's maximum debt will increase from $231,200 to $404,000, which should also be taken in consideration. These numbers are magnified when the contract is in millions or hundreds of millions of dollars and when the construction time is much longer than six months.

Appendix B
CSI MasterFormat

CSI MasterFormat 1995

Construction Products and Activities

Division 1—General Requirements

Division 2—Site Construction

Division 3—Concrete

Division 4—Masonry

Division 5—Metals

Division 6—Wood and Plastics

Division 7—Thermal and Moisture Protection

Division 8—Doors and Windows

Division 9—Finishes

Division 10—Specialties

Division 11—Equipment

Division 12—Furnishings

Division 13—Special Construction

Division 14—Conveying Systems

Division 15—Mechanical

Division 16—Electrical

CSI MasterFormat 2016[1]

PROCUREMENT AND CONTRACTING REQUIREMENTS GROUP

Division 00 Procurement and Contracting Requirements

SPECIFICATIONS GROUP

---

1. MasterFormat 2016, The Construction Specifications Institute, April 2016.

GENERAL REQUIREMENTS SUBGROUP
Division 01 General Requirements

FACILITY CONSTRUCTION SUBGROUP
Division 02 Existing Conditions

Division 03 Concrete

Division 04 Masonry

Division 05 Metals

Division 06 Wood, Plastics, and Composites

Division 07 Thermal and Moisture Protection

Division 08 Openings

Division 09 Finishes

Division 10 Specialties

Division 11 Equipment

Division 12 Furnishings

Division 13 Special Construction

Division 14 Conveying Equipment

Divisions 15–19 Reserved

FACILITY SERVICES SUBGROUP
Division 20 Reserved

Division 21 Fire Suppression

Division 22 Plumbing

Division 23 Heating, Ventilating, and Air Conditioning

Division 24 Reserved

Division 25 Integrated Automation

Division 26 Electrical

Division 27 Communications

Division 28 Electronic Safety and Security

Division 29 Reserved

SITE AND INFRASTRUCTURE SUBGROUP
Division 30 Reserved

Division 31 Earthwork

Division 32 Exterior Improvements

Division 33 Utilities

Division 34 Transportation

Division 35 Waterway and Marine Construction

Divisions 36–39 Reserved

PROCESS EQUIPMENT SUBGROUP

Division 40 Process Interconnections

Division 41 Material Processing and Handling Equipment

Division 42 Process Heating, Cooling, and Drying Equipment

Division 43 Process Gas and Liquid Handling, Purification, and Storage Equipment

Division 44 Pollution and Waste Control Equipment

Division 45 Industry-Specific Manufacturing Equipment

Division 46 Water and Wastewater Equipment

Division 47 Reserved

Division 48 Electrical Power Generation

Division 49 Reserved

Appendix C

Sample Estimating Forms

ESTIMATE SHEET

Estimator: Project:

Checked by: Date:

Problem Description:

Assumptions:

Quantities:

Item	Description	Each	Dimensions			Total Quantity	Unit	Means Item	Unit Bare Cost			
			L'	W'	D"				Materials	Labor	Equip.	Total
1	3-2X12 Girder	3	32	2	12	0.192	MBF	06 11 1010 5060	$680.00	$242.00	$0.00	$922.00
2	2X10 Joists	54	12	2	10	1.080	MBF	06 11 1018 2720	$670.00	$460.00	$0.00	$1,130.00
3	2X10 Band Joists	8	8	2	10	0.107	MBF	06 11 1018 2720	$670.00	$460.00	$0.00	$1,130.00
4	1X3 Bridging	48	NA	NA	NA	0.480	C Pr	06 11 1006 0011	$51.00	$265.00	$0.00	$316.00
5	~TQT"subfloor plywood CDX	1	32	24	NA	768	SF	06 16 2310 0200	$0.82	$0.55	$0.00	$1.37

| Unit Cost Incl O & P | CCI | | WC % | Avg Fixed OH | OH | Profit | Labor Add-ons | Total Bare Cost | Total Cost incl. OH & P | |
	Mat.	Inst. Total							method 1	method 2
$1,125.00	109.80%	119.00% 115.10%	21.58%	16.30%	11%	10%	58.88%	$198.65	$248.62	$245.54
$1,450.00	109.80%	119.00% 115.10%	21.58%	16.30%	11%	10%	58.88%	$1,385.70	$1,802.47	$1,813.00
$1,300.00	109.80%	119.00% 115.10%	21.58%	16.30%	11%	10%	58.88%	$136.86	$159.61	$179.09
$460.00	109.80%	119.00% 115.10%	21.58%	16.30%	11%	10%	58.88%	$178.25	$254.14	$270.06
$1.75	109.80%	119.00% 115.10%	21.58%	16.30%	11%	10%	58.88%	$1,194.13	$1,546.94	$1,559.24
							Total	$3,094.00	$4,012.00	$4,067.00

QUANTITY SHEET

	SHEET NO.
PROJECT	ESTIMATE NO.
LOCATION ARCHITECT	DATE
TAKE OFF BY EXTENSIONS BY:	CHECKED BY:

DESCRIPTION	NO.	DIMENSIONS				UNIT		UNIT		UNIT		UNIT

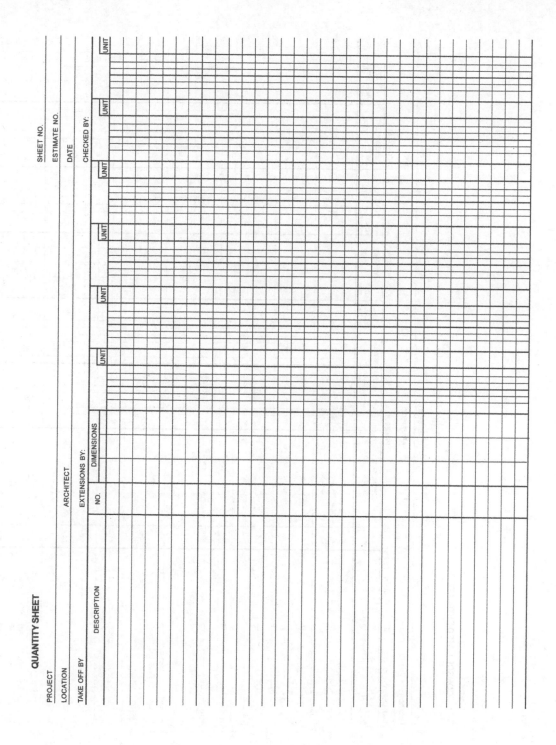

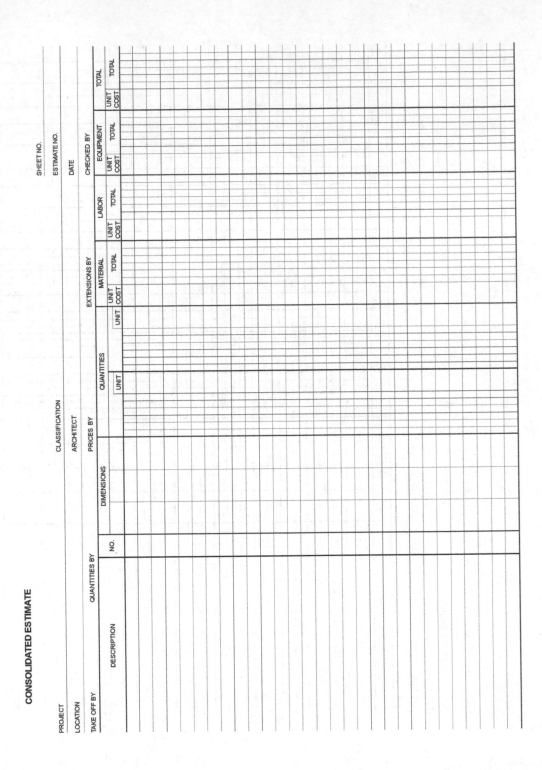

PROJECT OVERHEAD SUMMARY

PROJECT		SHEET NO.
		ESTIMATE NO.
LOCATION	ARCHITECT	DATE
QUANTITIES BY:	PRICES BY:	EXTENSIONS BY:	CHECKED BY:

DESCRIPTION	QUANTITY	UNIT	MATERIAL/EQUIPMENT		LABOR		TOTAL COST	
			UNIT	TOTAL	UNIT	TOTAL	UNIT	TOTAL
Job Organization: Superintendent								
Project Manager								
Timekeeper & Material Clerk								
Clerical								
Safety, Watchman & First Aid								
Travel Expense: Superintendent								
Project Manager								
Engineering: Layout								
Inspection/Quantities								
Drawings								
CPM Schedule								
Testing: Soil								
Materials								
Structural								
Equipment: Cranes								
Concrete Pump, Conveyor, Etc.								
Elevators, Hoists								
Freight & Hauling								
Loading, Unloading, Erecting, Etc.								
Maintenance								
Pumping								
Scaffolding								
Small Power Equipment/Tools								
Field Offices: Job Office								
Architect/Owner's Office								
Temporary Telephones								
Utilities								
Temporary Toilets								
Storage Areas & Sheds								
Temporary Utilities: Heat								
Light & Power								
Water								
PAGE TOTALS								

Appendix D
References

AACE International website, www.aacei.org.

AACE International. *Recommended Practice No. 10S-90, Cost Engineering Terminology*. Morgantown, WV: AACE International, November 14, 2014, revised October 31, 2017. http://web.aacei.org/docs/default-source/rps/10s-90.pdf?sfvrsn=16.

AACE International. *Recommended Practice No. 33R-15, Developing the Project Work Breakdown Structure*. Morgantown, WV: AACE International, March 1, 2016.

AACE International. *Recommended Practice No. 35R-09, Development of Cost Estimate Plans: As Applied for the Building and General Construction Industries*. Morgantown, WV: AACE International, February 14, 2015.

AACE International. *Recommended Practice No. 11R-88, Required Skills and Knowledge of Cost Engineering*. Morgantown, WV: AACE International, Rev. June 18, 2013.

AACE International. *Recommended Practice No. 18R-97, Cost Estimate Classification System: As Applied in Engineering, Procurement, and Construction for the Process Industries*. Morgantown, WV: AACE International, revised March 1, 2016. https://web.aacei.org/docs/default-source/toc/toc_18r-97.pdf?sfvrsn=4.

AACE International. *Recommended Practice No. 56R-08, Cost Estimate Classification System—As Applied in the Building and General Construction Industries*. Morgantown, WV: AACE

International, Rev. December 5, 2012. https://web.aacei.org/docs/default-source/toc/toc_56r-08.pdf?sfvrsn=4.

AACE International. *Recommended Practice No. 58R-10, Escalation Principles and Methods Using Indices*. Morgantown, WV: AACE International, May 25, 2011.

Adrian, James J. *Construction Estimating: An Accounting and Productivity Approach*. Champaign, IL: Stipes Publishing, 1993.

American Institute of Timber Construction. *Timber Construction Manual* (6th ed.). Hoboken, NJ: John Wiley and Sons, 2012.

American Society of Professional Estimators. *Standard Estimating Practice* (9th ed.). Vista, CA: BNi Publications, 2014.

Amos, Scott J. (ed.). *Skills & Knowledge of Cost Engineering* (5th ed.). Morgantown, WV: AACE International, 2011.

Arthur, William B. *The New Building Estimator: A Practical Guide to Estimating the Cost of Labor and Material in Building Construction, from Excavation to Finish*. Charleston, SC: Nabu Press, 2014. Originally published by the David Williams Company, New York, 1909.

Ashworth, Allan. *Pre-contract Studies: Development Economics, Tendering and Estimating* (3rd ed.). Oxford, UK: Wiley-Blackwell, 2008.

Au, Tung, and Thomas P. Au. *Engineering Economics for Capital Investment Analysis* (2nd ed.). Upper Saddle River, NJ: Prentice Hall, 1991.

Au, Tung, and C. Hendrickson. *Project Management for Construction*. Upper Saddle River, NJ: Prentice Hall, 1989.

Bartholomew, Stuart H. *Construction Contracting: Business and Legal Principles* (2nd ed.). Upper Saddle River, NJ: Pearson, 2001.

Benedict, Bert, and Gordon Anderson. *Estimating for Residential & Commercial Construction*. Clifton Park, NY: Delmar Publishing, 1994.

Blank, L. T., and A. J. Tarquin. *Engineering Economy* (7th ed.). New York: McGraw-Hill, 2011.

Bledsoe, John D. *Successful Estimating Methods*. Kingston, MA: RSMeans, 1992.

Clark, F. D., and A. B. Lorenzoni. *Applied Cost Engineering* (3rd ed.). New York: Marcel Dekker, 1997.

Clough, R. H., and G. A. Sears. *Construction Contracting: A Practical Guide to Company Management* (7th ed.). Hoboken, NJ: John Wiley and Sons, 2005.

Collier, C. A., and C. A. Glagola. *Engineering Economic and Cost Analysis* (3rd ed.). Upper Saddle River, NJ: Pearson, 1998.

Construction Industry Institute. *Report RS257-1—Global Procurement and Materials Management*, www.construction-institute.org/scriptcontent/more/257_1_more.cfm. Austin, TX: Construction Industry Institute.

Cook, Paul J. *Quantity Takeoff for Contractors*. Kingston, MA: RSMeans, 1989.

DelPico, Wayne J. *Estimating Building Costs for the Residential & Light Commercial Contractor* (2nd ed.). Kingston, MA: RSMeans, 2012.

Diamant, Leo, and C. R. Tumblin. *Construction Cost Estimating* (2nd ed.). New York: John Wiley and Sons, 1990.

Fatzinger, James A. S. *Basic Estimating for Construction* (2nd ed.). Upper Saddle River, NJ: Pearson, 2003.

Gould, Frederick E. *Managing the Construction Process* (4th ed.). Upper Saddle River, NJ: Pearson, 2011.

Grant, E. L., W. G. Ireson, and R. S. Leavenworth. *Principles of Engineering Economy* (8th ed.). New York: John Wiley and Sons, 1990.

Helton, Joseph E. *Simplified Estimating for Builders and Engineers* (2nd ed.). Upper Saddle River, NJ: Prentice Hall, 1992.

Herbsman, Zohar J. "Lane Rental—Innovative Way to Reduce Road Construction Time." *Journal of Construction Engineering and Management* 124(5) (September/October 1998): 411–417.

Hollmann, John K. (ed.). *Total Cost Management Framework: An Integrated Approach to Portfolio, Program and Project Management.* Morgantown, WV: AACE International, 2012.

Holm, L., J. E. Schaufelberger, D. Griffin, and T. Cole. *Construction Cost Estimating: Process and Practices.* Upper Saddle River, NJ: Prentice Hall, 2004.

Humphreys, Kenneth K., and P. Wellman. *Basic Cost Engineering.* New York: Marcel Dekker, 1996.

Humphreys, Kenneth K. *Project and Cost Engineers' Handbook* (4th ed.). Boca Raton, FL: CRC Press, 2004.

Hurd, M.K. *Formwork for Concrete* (7th ed.). Farmington Hills, MI: American Concrete Institute, 2005.

Jackson, Barbara J. Construction Management Jumpstart (2nd ed.). Hoboken, NJ: John Wiley and Sons, 2010.

Jackson, I. J. *Financial Management for Contractors* (4th ed.). Tampa, FL: FMI Corporation, 2002.

Jelen, F. C., and J. H. Black. *Cost and Optimization Engineering* (2nd ed.). New York: McGraw-Hill, 1983.

Kitchens, Michael. *Estimating and Project Management for Building Contractors.* Reston, VA: ASCE Press, 1996.

March, Chris. *Finance and Control for Construction.* New York: Taylor & Francis, 2009.

Mordue, S., P. Swaddle, and D. Philp. *Building Information Modeling for Dummies.* Chichester, U.K.: John Wiley & Sons Ltd., 2016.

Mubarak, Saleh. *Construction Project Scheduling & Control* (4th ed.). Hoboken, NJ: John Wiley and Sons, 2019.

Neil, James M. *Construction Cost Estimating for Project Control*. Englewood Cliffs, NJ: Prentice Hall, 1982.

Newnan, D. G., T. Eschenbach, and J. Lavelle. *Engineering Economic Analysis* (13th ed.). New York: Oxford University Press, 2017.

O'Brien, James J. *Preconstruction Estimates*. New York: McGraw-Hill, 1994.

Ostwald, Phillip F., and Timothy S. McLaren. *Cost Analysis and Estimating for Engineering and Management*. Upper Saddle River, NJ: Pearson, 2003.

Park, William R., and Wayne B. Chapin. *Construction Bidding* (2nd ed.). New York: John Wiley and Sons, 1992.

Peterson, Steven J., and Frank R. Dagostino. *Estimating in Building Construction* (8th ed.). Upper Saddle River, NJ: Pearson, 2014.

Peurifoy, R. L. *Estimating Construction Costs* (6th ed.). New York: McGraw-Hill, 2013.

Pierce, David R., Jr. *Project Scheduling and Management for Construction* (4th ed.). Kingston, MA: RSMeans, 2013.

Plan Reading and Material Takeoff. Kingston, MA: RSMeans, 1994.

Pratt, David. *Fundamentals of Construction Estimating* (3rd ed.). Clifton Park, NY: Delmar Publishing, 2010.

Pray, Richard. *2014 National Construction Estimator*. Carlsbad, CA: Craftsman Book Company, 2013.

Project Management Institute, *A Guide to the Project Management Body of Knowledge (PMBOK® Guide)* (5th ed.). Newton Square, PA: Project Management Institute, Inc., 2013.

Riggs, Henry E. *Financial and Economic Analysis for Engineering and Technology Management* (2nd ed.). Hoboken, NJ: Wiley-Interscience, 2004.

RSMeans Assemblies Cost Data. Kingston, MA: RSMeans (annual).

RSMeans Building Construction Cost Data. Kingston, MA: RSMeans (annual).

RSMeans Estimating Handbook (3rd ed.). Kingston, MA: RSMeans, 2009.

RSMeans Illustrated Construction Dictionary. Kingston, MA: RS Means, 2009.

RSMeans Square Foot Costs. Kingston, MA: RSMeans (annual).

Riggs, James L., David D. Bedworth, and Sabah U. Randhawa. *Engineering Economics* (4th ed.). New York: McGraw-Hill, 1996.

Schuette, S. D., and R. W. Liska. *Building Construction Estimating.* New York: McGraw-Hill, 1994.

Shim, J. K., and N. Henteleff. *What Every Engineer Should Know About Accounting and Finance.* New York: Marcel Dekker, 1995.

Simmons, H. L. *Olin's Construction: Principles, Materials, and Methods* (9th ed.). Hoboken, NJ: John Wiley and Sons, 2011.

Sundberg, Elmer W., and Thomas E. Proctor. *Building Trades Printreading Parts 1 and* 2. Orland Park, IL: American Technical Publishers, 1987.

Thomas, Paul I. *The Contractors Field Guide.* Upper Saddle River, NJ: Prentice Hall, 1991.

Ward, Sol A. *Cost Engineering for Effective Project Control.* New York: John Wiley and Sons, 1992.

Index

Page numbers followed by *f* and *t* refer to figures and tables, respectively.